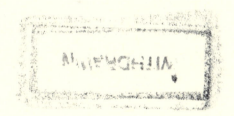

High Performance
Liquid Chromatography

Fundamental Principles
and Practice

High Performance Liquid Chromatography

Fundamental Principles and Practice

Edited by

W.J. LOUGH
School of Health Sciences
University of Sunderland

and

I.W. WAINER
Department of Oncology
McGill University
Montreal

BLACKIE ACADEMIC & PROFESSIONAL
An Imprint of Chapman & Hall

London · Glasgow · Weinheim · New York · Tokyo · Melbourne · Madras

Published by
Blackie Academic & Professional, an imprint of Chapman & Hall,
Wester Cleddens Road, Bishopbriggs, Glasgow G64 2NZ

Chapman & Hall, 2–6 Boundary Row, London SE1 8HN, UK

Blackie Academic & Professional, Wester Cleddens Road, Bishopbriggs, Glasgow G64 2NZ, UK

Chapman & Hall GmbH, Pappelallee 3, 69469 Weinheim, Germany

Chapman & Hall USA, 115 Fifth Avenue, New York, NY 10003, USA

Chapman & Hall Japan, ITP-Japan, Kyowa Building, 3F, 2-2-1 Hirakawacho, Chiyoda-ku, Tokyo 102, Japan

DA Book (Aust.) Pty Ltd, 648 Whitehorse Road, Mitcham 3132, Victoria, Australia

Chapman & Hall India, R. Seshadri, 32 Second Main Road, CIT East, Madras 600 035, India

First edition 1995

© 1996 Chapman & Hall

Typeset in 10/12pt Times by Acorn Bookwork, Salisbury, Wiltshire
Printed and bound in Great Britain by Hartnolls Limited, Bodmin, Cornwall

ISBN 0 7514 0076 9

A catalogue record for this book is available from the British Library
Library of Congress Catalog Card Number: 95-80811

Preface

Since its infancy in the late 1960s and early 1970s, the use of high performance liquid chromatography (HPLC, or LC as it is increasingly becoming known) has expanded rapidly. Today, it is a highly popular analytical technique which is extensively used in many fields of activity. Not surprisingly, it has been the inspiration for much scientific literature, including a significant number of books. These books have taken various forms, such as practical handbooks, modified short-term manuals, reports on selected topics, theoretical treatises and even attempts at comprehensive works.

In almost every way, HPLC may be regarded as a mature technique. However, there is one sense in which it is still developing. Only recently has its importance been acknowledged by accepting it as a subject worthy of inclusion in undergraduate science courses. The aim of this book is therefore to support the newly acquired status of HPLC by providing a presentation of the technique primarily aimed at undergraduate students on courses containing a significant component of analytical science. The intention has been to come up with a different slant to that already available, characterised by an emphasis on understanding. Why is HPLC so useful? When should it be used? What is the reason for certain practices? Most importantly, how does HPLC work? The responsibility to deliver a thorough coverage of theory has not been shirked.

While the book is, as the editors intended, for the teaching of undergraduate analytical chemistry, another aim is to make the book as versatile as possible. HPLC is taught not only on chemistry-based courses, but also on non-chemistry applied science courses. This is particularly the case for taught Masters courses. Students who have taken a more traditional science first degree often enhance their job prospects by proceeding to an applied science Masters degree, many of which have a strong analytical content. Versatility has therefore been sought by making this a contributed rather than a single author text. This is particularly important for the applications chapters, since an expert in each application area is best able to identify those aspects of HPLC which are important in that application area. Hence the analytical student will obtain an accurate view of the role played by HPLC in, say, environmental science, and the environmental science student will be directed to the most relevant parts of the book. By allowing a certain amount of overlap between contributors, versatility should also arise, because each chapter is more able to stand on its own and therefore be more accessible.

We wish to thank our colleagues and associates who have taken part in this project and trust that our readers will find the book fulfils the rationale described above.

W.J.L.
I.W.W.

Contributors

P. Hambleton   School of Health Sciences, University of Sunderland, Fleming Building, Chester Road, Sunderland SR1 3SD, UK

A.J. Handley   ICI Chemicals and Polymers Ltd, PO Box 14, The Heath, Runcorn, Cheshire WA7 4QF, UK

D.K. Lloyd   Department of Oncology, McGill University, 3626 St Urbain Street, Montreal, Quebec H2X 2P2, Canada

W.J. Lough   School of Health Sciences, University of Sunderland, Fleming Building, Chester Road, Sunderland SR1 3SD, UK

T. Noctor   Department of Biopharmaceutical Analysis, Hazleton Europe, Otley Road, Harrogate, North Yorkshire HG3 1PY, UK

D. Perrett   Department of Medicine, St Bartholomew's Hospital Medical College, West Smithfield, London EC1A 7BE, UK

P.J. Rennie   Yorkshire Environmental LabServices, Templeborough House, Mill Close, Rotherham, South Yorkshire S60 1BZ, UK

C.M. Riley   Du Pont Merck Pharmaceutical Company, PO Box 80400, Wilmington, DE 19880–0400, USA

D. Stevenson   Robens Institute of Health and Safety, University of Surrey, Guildford, Surrey GU2 5XH, UK

I.W. Wainer   Department of Oncology, McGill University, Montreal General Hospital, 1650 Cedar Avenue, Montreal, Quebec, H3G 1A4, Canada

P. White   Forensic Science Unit, University of Strathclyde, Royal College, 204 George Street, Glasgow G1 1XW, UK

Contents

4   Support materials and solvents   79
P. HAMBLETON

5   Instrumentation: pumps, injectors and column design   97
T. NOCTOR

6   Instrumentation: detectors and integrators   114
D.K. LLOYD

7   Method development and quantitation   143
W.J. LOUGH and I.W. WAINER

11   Environmental analysis   234
P.J. RENNIE

12   Food, organic and pharmaceutical applications   248
W.J. LOUGH and I.W. WAINER

Index   270

1 Introduction

W.J. LOUGH and I.W. WAINER

1.1 Analysis and chromatography

Before embarking on a study of high-performance liquid chromatography (HPLC), it is instructive to take a look at how this analytical technique emerged and to consider its relative importance in the field of analytical chemistry. Although HPLC is used to determine what is in a sample (qualitative analysis), its primary application is as a quantitative analytical tool and as such it plays a key role in analytical methods, since the analytical technique is only one part of the overall analytical method. An analytical method may be thought of as consisting of five distinct parts:

- defining the problem
- taking the sample
- sample pre-treatment
- measurement
- calculating, assessing and reporting the results

While great importance is attached to the measurement step, the other parts of an analytical method are by no means trivial. For example, all the subsequent steps depend on defining the problem. In choosing the appropriate subsequent steps in the method it must be borne in mind exactly what it is that it is wished to find out about the sample. For example, it is very likely that the steps taken to confirm that the content of a formulated drug substance was within specification (i.e. the amount of drug present was within acceptable limits as stated on the container label) would differ from those taken to determine a minor degradation product. Similarly, there are large differences in the sampling of liquids, gases and solids. Liquids and gases are homogenous and sampling often only amounts to ensuring that a large enough sample has been taken. Solid samples on the other hand can be non-homogenous and it is very important that great care is taken to ensure that a representative sample is taken by sampling from different parts of the whole. 'Sample pre-treatment' also may be a simple and straightforward task amounting only to ensuring that the analyte is in a phase compatible with the measurement technique and present in an appropriate concentration. However, as will be seen elsewhere in this book, sample pre-treatment is often a very complex and important operation. The importance of 'calculating and reporting the results' is self-evident and obviously cannot be emphasised enough. The assessment of the results particularly by the extensive use of statistical methods is also an important and ever-increasing feature of the work of an analyst.

To understand why HPLC is frequently used in quantitative analytical methods, it is useful to assess whether or not an analytical method is suitable for its intended purpose and, in doing so, consider the deficiencies in methods employing classical measurement steps. Analytical method validation is the process of assessing the fitness for purpose of an analytical method; in choosing an analytical method issues such as cost, simplicity, operator experience, availability etc. are of secondary importance to the actual validity of the method under consideration. In the validation procedure, tests are typically carried out for the following properties:

Specificity.   An assay is specific if the 'analytical response' (i.e. that which is measured) arises from the analyte of interest only and cannot arise from any other compound likely to be present in the sample.

Robustness.   A robust or rugged method is one which is not greatly affected by minor changes in experimental variables that might reasonably be expected during the course of an assay.

Linearity.   A method is linear if there is a linear relationship between the 'analytical response' and concentration of analyte in the sample solution over a specified range of concentrations of the analyte (e.g. 20–120% of the anticipated levels of analyte concentration in the sample solutions). It would also be anticipated that the plot of 'analytical response' versus analyte concentration would have a negligible intercept.

Precision.   The precision of a method is the degree of scatter of the results and is usually reported as a percentage relative standard deviation. It is often subdivided into repeatability (precision on replicate measurements of the same solution) and reproducibility (precision of the results from measurements of different solutions, i.e. of the complete method).

Accuracy.   An assay is accurate if the mean result from the assay is the same as the true value (i.e. there is no bias (systematic error)). Assessment is notoriously difficult since often the true value is not known but good accuracy is generally a consequence of the other validation parameters being within acceptable limits.

Limit of detection.   The limit of detection is the amount of analyte which can be reliably detected under the stated experimental conditions. Often a statistical approach may be adopted in defining what is 'reliable' and what is not.

Limit of quantitation. The limit of quantitation is the amount of analyte which can be reliably quantified under the stated experimental conditions. Often a statistical approach may be adopted in defining what is 'reliable' and what is not.

Stability in solution. It is necessary to study stability in solution in order to be able to make a statement such as, e.g. "the analyte was sufficiently stable in solution in the solvent used for preparing sample solutions for reliable analysis to be carried out."

1.2 HPLC versus other analytical methods

The subject of analytical method validation is very complex but the brief explanation given above of the tests that must be carried out should be sufficient to enable an evaluation of the relatively simple techniques that were used prior to the development of chromatographic techniques such as HPLC.

1.2.1 Volumetric analysis

Volumetric analysis using titration methods is very cheap and simple. In recent times it has even been possible to automate such methods thereby reducing operator errors and increasing sample throughput. However, although such methods may be adapted to give reasonably low limits of detection and quantitation they are normally used in the determination of millimolar to high micromolar quantities in fairly concentrated solutions (e.g. 0.05 M). Precision, accuracy and robustness are often problems in volumetric analysis but only if there has been an inappropriate choice of indicator. Stability in solution is related to the analyte and the solvent in which it is dissolved and is not dependent on the measurement technique being used. Probably the most important validation test is that for specificity and unfortunately it is here that volumetric analysis falls down. This is particularly the case for determining the purity of complex molecules such as drugs. Such molecules are generally prepared by a multi-stage synthesis and the drug sample may contain water, solvents, trace catalysts, inorganics, and structurally related synthetic by-products, i.e. intermediates, isomers, degradation products and products of side reactions. For a basic compound it is highly likely that all or almost all of these structurally related impurities will also be basic. Therefore titration against an acid would not give purity but would give some measure of the total amount of drug, structurally related impurities and other basic impurities. Thinking particularly of this example of drugs, this is obviously not an acceptable situation, since it is usually most important to determine the structurally related impurities since it is these that are most likely to give rise to toxicological problems.

1.2.2   Ultraviolet spectrophotometry

Similar difficulties arise when using ultraviolet (UV) spectrophotometry, in which analyte concentrations are determined by measuring the absorbance of UV light by the sample solution. For this common, simple technique, detectability (down to low nanogram levels) is good and there is a wide linear range. However, that all-important feature, specificity, is again a problem. For structurally related impurities of a UV-absorbing compound it is highly likely that they will contain a very similar chromophore (i.e. the part of the molecule which actually absorbs the UV radiation) or part of the chromophore to that present in the compound. Therefore incident radiation at any particular wavelength of the UV spectrum of the compound would almost certainly be absorbed by structurally related compounds as well as the compound itself.

1.2.3   Using chromatography to improve specificity

There are two approaches to solve this critical problem of lack of specificity in the relatively simple analytical techniques discussed above. One is to measure a physical or chemical property of the analyte that is unique. This is a tall order since not just acidity, basicity and UV spectrum but almost all the properties of compounds are similar to those of their structurally related impurities. The property that perhaps comes closest to being unique is the mass spectrum of a compound (i.e. the pattern obtained when a compound is ionised under high energy conditions and the resultant ions traverse a magnetic field) which gives characteristic information on molecular mass and the mass of fragments of the molecule. However, mass spectrometry (MS) requires expensive instrumentation and therefore only under very challenging circumstances would it be considered an appropriate option for routine quantitative analysis (also it can fail to distinguish between isomers).

The second approach to specificity in analysis is fortunately more readily achievable. It involves separating all the components of a mixture from one another (or at least from the analyte) before measurement takes place. To a certain extent this may be done in the sample pre-treatment step of an analytical method which, as well as ensuring the analyte is in a suitable phase in a suitable concentration for measurement, typically involves some form of sample 'clean-up'. By using simple separation processes such as dissolution and filtration, liquid–liquid extraction, etc. in the sample pre-treatment step of the analytical method, it is possible to rid the sample of gross interferences. To use a pharmaceutical illustration again, in determining drugs in biological fluids, it is possible to remove large and small polar molecules endogenous to the biological fluid leaving behind primarily the drug and its metabolites. However, such sample pre-treatment will rarely isolate the analyte from structurally related compounds present in the sample. It is therefore necessary to use more powerful separation processes, i.e. chromatographic techniques.

Chromatography is a differential migration process where sample components are distributed between a stationary and a mobile phase. Depending on the affinity of a component for either or both of these phases, the rate of migration varies from zero to the velocity of the mobile phase. The distribution coefficient (D) for an analyte between the two phases is

$$D = A_s / A_m$$

where A_s is the activity (often approximates to concentration) of the analyte in the stationary phase and A_m is the activity of analyte in the mobile phase. The larger the value for D the greater the affinity of the analyte for the stationary phase. If two components of a mixture have even a very small difference in their D values, they may be separated, since the chromatographic process of separation may be repeated a large number of times and this serves to amplify that difference as migration proceeds.

Since chromatographic techniques are separation processes they could be regarded as very powerful means for 'sample pre-treatment'. However, since the term, chromatography, is generally regarded as incorporating the 'measurement' step (i.e. detection subsequent to the separation), it is more accurate to view it as a combination of 'sample pre-treatment' and 'measurement'. Also it is frequently not the complete 'sample pre-treatment' since very often a more rudimentary form of sample pre-treatment will be required prior to chromatography.

HPLC is but one of several chromatographic techniques. Having looked at the role of chromatographic techniques in analysis in order to appreciate their importance, it now remains to look at how HPLC emerged as a logical progression out of the historical development of chromatography.

1.3 Historical development of chromatography

There are a number of examples of work carried out prior to the twentieth century in which experiments were conducted which are recognisable in retrospect as early forms of chromatography. However, it was the work of Russian botanist Michael Tswett, first publicly reported in 1903, that constituted the first systematic study of what would today be recognised as chromatography. He used column liquid chromatography in which the stationary phase was a solid adsorbent packed into a glass column and the mobile phase was a liquid. He carried out experiments on chlorophyll extracts in petroleum spirit with over 100 adsorbents. Although most of these adsorbents are now no longer important, it is interesting to note in the list the inclusion of materials such as silica, alumina, charcoal, calcium carbonate, magnesia and sucrose which are still in use. He also confirmed the identities of the fractions obtained by spectrophotometry at various wavelengths thus anticipating the commonest mode of detection in liquid chromatography. In a paper published in 1906 he first used the term 'chromatography' (from the Greek *khroma, -atos* colour; *graphos* writing). In 1910 Tswett obtained a Russian doctorate (having received his earlier education in Switzerland) and his doctoral thesis was published as a monograph again showing further development and refinement of his ideas. This monograph marked the end of his chromatographic work. This was not surprising since he was a botanist and chromatography was merely a means to an end.

Despite the fact that Tswett travelled widely in Germany, France and Holland and must have discussed his ideas on chromatography with leading scientists of those countries, chromatographic methods were largely ignored until the early 1930s. One of the few exceptions was the work of an American, L.S. Palmer, who from 1913 onwards published a number of papers describing the use of column liquid chromatography for the separation of the pigments in plants and dairy products. This culminated in a book published in 1922 summarising his work and in which the earlier work of Tswett was acknowledged. There are various reasons for the lack of interest in chromatography at the time, the main one being that it was a relatively small-scale physical method and the scientific community was mainly orientated towards large-scale synthetic organic chemistry.

In 1930 in Germany, Edgar Lederer drew upon the work of Tswett and Palmer in using chromatography in an investigation into the pigments in egg yolk. Because of the relative speed of the technique it was possible to avoid the degradation of the carotene molecules. Thereafter there was steady success including the development of forms of chromatography other than column liquid chromatography and of instrumental methods of analysis (e.g. infrared spectroscopy and mass spectrometry) which would much later be incorporated into instrumental chromatography. In 1938 Eastern European workers carried out planar chromatography in which the powder was spread on a glass plate. Thin-layer chromatography had its origins in this work but at this time the plate had to be horizontal otherwise the layer of powder would be displaced.

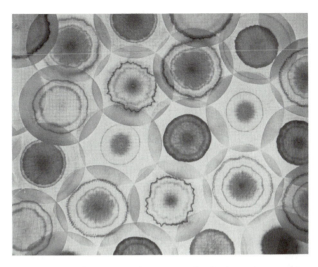

Figure 1.1 Paper chromatography. Even today lower school science lessons often make use of paper chromatography to separate the colours of Smarties, for example. Reprinted with permission (J. Nicholson, *Chemistry in Britain* **30** (1994) 658).

The major breakthrough that would eventually lead to many of the developments in modern chromatography came in 1941 with the work of Martin and Synge. They carried out partition chromatography of amino acids using silica wetted with water and treated with an indicator. Later Martin and other workers in 1943 used cellulose, i.e. paper chromatography (Figure 1.1). However, what was much more important was that Martin and Synge produced the first mathematical treatment of chromatographic theory (for which they won the Nobel Prize in 1952) using plate theory and predicted many of the developments in chromatography that were later to become possible.

For example it was not until 1952 that Martin, with co-worker James, was able to put into practice gas–liquid chromatography, in which volatile compounds are analysed using a gaseous mobile phase and a stationary phase consisting of an involatile liquid coated on a solid support. While it is true that chemistry only advances as analytical techniques advance, what is more relevant is that analytical techniques advance as there is an economic or even military necessity for them to improve and with it the required ancillary technologies. There had not been much dissemination of scientific information during World War II but looking back it is clear that analysis had moved forward, for example because of the need to examine aviation fuels. Similarly, around the time of Martin and James' work in 1952, the petroleum industry was expanding rapidly to meet the demands of post-war economic development including the early stages of the proliferation of motor transport. Therefore the petroleum industry in

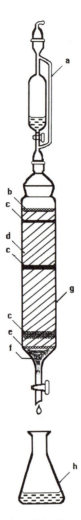

Figure 1.2 Open-column chromatography. (a) Pressure equalised dropping funnel with optional air or nitrogen over pressure. (b) Narrow bed of coarse sand or glass wool to protect top of the bed of packing material. (c) Bands of different solutes migrate down the column at different rates depending on the strength of interactions with the packing material and widening as they traverse the bed. (d) Bed of packing material, typically porous irregular silica particles ($> 100 \mu m$). (e) Sinter to retain bed of packing material in column. (f) Glass wool. (g) Glass column; there can be great variations in dimensions. (h) Conical flask(s) to collect eluent from the column; in early use of this technique the colour of the eluent was used to detect when a separated band was eluting from the end of the column.

particular took up gas chromatography with great enthusiasm and poured a considerable effort into its development. The reasons for this are not difficult to understand. Hydrocarbon mixtures are composed of a large number of chemically very similar and very unreactive compounds which are difficult to separate by conventional means. Even in the early 1950s research on the composition of crude oil and oil products made considerable use of column liquid chromatography but the equipment and technique were still in the Tswett era (Figure 1.2). Hydrocarbon mixtures could be separated into saturates and mono-, di- and poly-aromatics by a combination of adsorbents. However, separation of individual compounds was, in general, not possible and the work was slow, labour intensive and therefore expensive. The possibility of being able to analyse complex hydrocarbon mixtures and automatically measure the amount of individual hydrocarbons present in about an hour was extremely attractive and offered the possibility of analysing mixtures which until then had been completely intractable.

By the mid-1960s the majority of the gas chromatography methods and equipment still in use had been described in principle. Also the limitations of the technique had become clear. Returning to the theme of economic necessity driving analytical science, there was a need, particularly in the increasingly important pharmaceutical and agrochemical industries, for chromatographic techniques by which involatile and thermally unstable compounds could be analysed without the need for chemical derivatisation.

In 1956 Stahl was responsible for the introduction of thin-layer chromatography (TLC). The stationary phase adsorbent was mixed with a binder which helped stick the separating layer to the glass plate. In this way the stationary phase is able to stand in a near to vertical position during the separation process while the mobile phase elutes up the plane of the stationary phase by capillary action (Figure 1.3). While TLC is suitable for the separation of involatile compounds, quantitation from the spots obtained on the plate is not easy. Accordingly, until recently, it has been regarded as merely a cheap and rapid screening method or as a method for obtaining fractions for further analysis.

During the late 1960s research was being conducted into making gas chromatography (GC) more applicable to involatile compounds by using a supercritical fluid instead of a gas as the mobile phase; supercritical fluids, such as carbon dioxide above its critical temperature of 31.1°C and critical pressure of 73.8 atm have much higher solvating power than gases. However, this turned out to be something of a false start for supercritical fluid chromatography (SFC) since at this time suitable equipment was not commercially available. Perhaps another factor was that the use of relatively large volumes of supercritical eluents such as pentane made the technique distinctly hazardous! The most important factor though was that the need for a chromatographic technique suitable for involatile compounds was satisfied more readily by the development of HPLC.

With developments in technology it was possible to apply chromatographic theory to the development of column liquid chromatography and

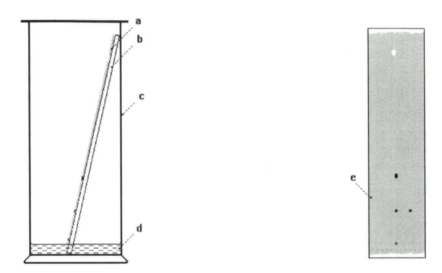

Figure 1.3 Thin-layer chromatography. (a) Thin-layer of adsorbent containing a binder so that it adheres to the glass plate. (b) Glass plate; the plates are made up by spreading a thick slurry of adsorbent on the plate, followed by drying in an oven; nowadays pre-prepared plate may be bought with the adsorbent on a plastic backing. (c) In the early days of TLC, gas jars were used but nowadays customised TLC tanks are available; the inside of the tank is lined with filter paper soaked in eluent in order that the atmosphere in the tank is saturated with eluent vapour. (d) Eluent; the plate is 'spotted' at a point which will be just above the level of the eluent. (e) The TLC plate. Many means of visualisation are possible (absorption of iodine, use of fluorescent agent on the adsorbent, concentrated sulphuric acid spray etc.); here a mixture is run against a reference standard of a key component in the mixture; however it is possible to run many samples simultaneously.

fulfil the predictions made many years earlier by Martin and Synge. The important improvement upon classical open-column liquid chromatography which came with HPLC was the use of very small particles for the solid adsorbent stationary phase. Because of this the bed of packing material had a much lower permeability so that it became necessary to use a pump to generate sufficient pressure to produce a fast enough flow rate. This gave rise to the improved technique being called High-Speed Liquid Chromatography and High-Pressure Liquid Chromatography. Soon these separate terms were replaced by the new term High-Performance Liquid Chromatography (HPLC), the new instrumental technique having better 'performance' in terms of resolving power, detection and quantitation as well as speed.

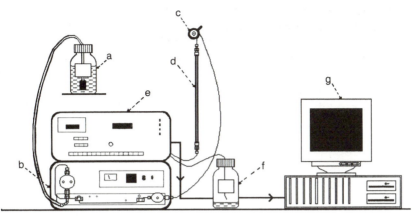

Figure 1.4 Modular HPLC system. (a) Labelled mobile phase reservoir; mobile phase is degassed and filtered. (b) Pump capable of delivering pulse-free flow at pressures up to 6000 psi. (c) Injection valve; in industrial applications an autosampler would normally be used. (d) Column; often with dimensions 250 × 4.6 mm i.d. (e) Detector; most popular means of detection is by UV absorbance. (f) Labelled waste mobile phase reservoir. (g) Integrator or PC-based data handling of signal from detector.

The use of high pressure brought with it major changes in the equipment required (Figure 1.4) compared to its traditional predecessor. The packing material had to be contained in a stainless-steel column, a pump which could operate at about 2000 psi and produce continuous pulse-free flow was required and special injection devices were required to apply sample solutions to the top of the column while it was under pressure. The other main difference was that instrumental detection systems were used rather than relying upon the visual observation of colour in the liquid eluting from the column. Also the scale of the separations was different since the technique was being used for quantitative analysis rather than for preparative purposes.

Not surprisingly HPLC proved to be much more suitable than gas chromatography for the analysis of pharmaceuticals, agrochemicals and other involatile substances important in key areas of industrial development at the time. As a result the technique underwent very rapid development. Some of the pioneers of HPLC at the time, to name just a few, were Huber, Kirkland, Knox, Snyder and Scott. Working principally from a theoretical standpoint, many advances were introduced in a short period of time. Notably during the 1970s, the fundamental problem of reducing the spreading of bands of the analytes as they traversed the HPLC column was tackled by moving from 40 μm pellicular particles (i.e. particles which consist of a non-porous core covered with a thin layer, typically about 5 μm thick) to 10 μm totally porous, irregularly shaped particles and then to 5 μm totally porous, spherical particles with a narrow particle size distribution. There was a simultaneous development in the understanding of solvent–support interactions. Better resolving power could be obtained in this way by appreciating the influence of experimental variables on selectivity and manipulating them accordingly. Exploitation of enhancements in selectivity to improve chromatographic resolution extended to the introduction of different modes of HPLC so that there are now a wide range of types of chemical equilibrium that have been used as vehicles for achieving a chromatographic separation. Instrument manufacturers also played an important role in the development of HPLC. One thinks today of some striking developments in automation and the use of computers in HPLC but before this phase of instrumental development, much work had to be done to ensure, above all, reliability. There were many important features of pumps, injectors and detectors which had to be addressed but the main focus was on the development of pumps which produced pulse-free continuous flow, injectors which could deliver small volumes (usually about 20 µl) repeatedly at high pressures and detectors which could give a large signal per unit concentration of analyte against a low background noise.

The end result of these developments and consequently the rationale behind them is covered in subsequent chapters in this book.

1.4 HPLC today

As referred to earlier, the emergence of HPLC was a logical progression in the development of chromatography. It is fairly safe to say that, because of its ready applicability to a wide range of sample types, by the mid-1970s it had become a pre-eminent chromatographic technique. This remains the position today.

However, other chromatographic techniques have also continued to develop and, given the different requirements of different application areas, they still have an important role to play. Even classical open-column chromatography has survived and is still used in synthetic organic chemistry laboratories because of its simplicity and low cost compared to other preparative techniques. Simplicity and low cost has also ensured that TLC has endured as a tool for monitoring the progress of organic reactions. TLC also has the advantages that many samples can be analysed simultaneously, and all components of a mixture may be visualised. Because of this it has an important role to play in areas such as forensic analysis, stability testing of drugs and analyses for which mass balance problems are found (i.e. when the purity of a sample by a relative method does not agree with the purity calculated on a 100% detected impurities basis). The scope of TLC has increased with the use of scanning densitometry to allow it to be more readily used for quantitation and with developments leading to the use of microparticles for high-performance TLC (HPTLC). However, these developments detract from the simplicity and low cost which is the main advantage of the technique.

Gas chromatography benefitted enormously from the introduction of open tubular GC in which the stationary phase was coated on the inner wall of a capillary rather on solid particles packed in a wider column. Although this innovation was described by Golay in 1957, it was not until the introduction of fused silica capillaries 20 years later that capillary GC gained wide acceptance. However, even with the high resolving power arising from the narrow band widths in capillary GC and the ready availability of a wide range of sensitive detectors, it is still less frequently used than HPLC in many important application areas such as pharmaceutical analysis. This is because its use is limited to the determination of analytes that may be readily volatilised without decomposition. The latter is less of a limitation for SFC which has benefitted from improved instrumentation for packed column work as well as the introduction of capillary SFC. However, despite something of a revival in the mid-1980s, SFC has on the whole not offered sufficient advantages for it to be used to replace HPLC. It is complementary to GC and HPLC and is used for a relatively narrow range of applications for which neither GC nor HPLC may be used without difficulty.

HPLC has been the most widely used chromatographic technique for some time now. It is estimated that currently, partly due to the enduring strength of important HPLC application areas such as the pharmaceutical industry during adverse economic conditions, that HPLC constitutes 60% of the worldwide separation science market while GC constitutes 35%. However, that is not to say that this position will be unchallenged in the future. Going back to the all-important need to obtain specificity in analytical methods, it can be seen that some emerging techniques can obtain specificity at least as effectively as HPLC. In capillary electrophoresis, in which analytes are separated by virtue of their differential migration through an electrolyte-filled capillary under the influence of an electric field, the very narrow band widths make it much easier to obtain separation and thereby specificity. Similarly high selectivity is readily, albeit expensively, obtained by high resolution mass spectrometers (in which analytes are ionised and separated by virtue of differential deflection as they traverse a magnetic field) or, more strikingly, when mass spectrometers are coupled in series (MS-MS). This may be regarded either as a highly effective separative technique or as a tuneable and therefore highly selective universal means of detection and measurement. While techniques such as capillary electrophoresis and MS-MS will be used in the future for many applications for which HPLC is now used, there is still no doubt that HPLC will continue to be a very widely used analytical technique.

Bibliography

Berezkin, V.G. (1990) *Chromatographic Adsorption Analysis. Selected Works of Mikhail Semenovitch Tswett*, Ellis Horwood, London.

Done, J.N., Kennedy, G.J. and Knox, J.H. (1972) *Nature* **237**, 77.

Ettre, L.S. (1975) *Anal. Chem.* **47**, 422A.

Ettre, L.S. (1990) *J. Chromatogr.* **535**, 3.

James, A.T. and Martin, A.J.P. (1952) *Biochem. J.* **50**, 679.

Martin, A.J.P. and Synge, R.L.M. (1941) *Biochem. J.* **35**, 1358.

Miller, J.C. and Miller, J.N. (1988) *Statistics for Analytical Chemists*, 2nd edn., Ellis Horwood, Chichester.

Nicholson, J. (1994) Paperwork made easy, *Chem. Br.* **30**, 658.

Sample Pre-Treatment, ACOL Series, Wiley, Chichester.

Smith, R.M., ed. (1988) *Supercritical Fluid Chromatography*, Royal Society of Chemistry, Cambridge.

Stahl, E. (1956) *Pharmazie* **11**, 633.

Tyson, J.F. (1989) Modern analytical chemistry *Anal. Proc.* **29**, 251–254.

Tyson, J.F. (1988) *Analysis – What Analytical Chemists Do*, Royal Society of Chemistry, London, 1988.

2 Efficiency, retention, selectivity and resolution in chromatography

C.M. RILEY

2.1 Introduction

Liquid chromatography (LC) is the term generally used to describe the separation of the components of a solution following differential migration of the solutes in a liquid flowing through a column packed with solid particles. It also applies to the more recent development of similar differential migration taking place in an open tube. Liquid chromatography is distinguished from the related techniques of gas chromatography (GC) and supercritical fluid chromatography (SFC) by the state of the mobile phase at the temperature and pressure of the system. Strictly speaking, thin-layer chromatography (TLC), paper chromatography (PC) and centrifugation chromatography (CC) are forms of liquid chromatography and the mechanisms of separation are the same. However, TLC and PC are generally termed planar chromatography because the separations are performed on stationary phases prepared as layers on flat surfaces.

The stationary phase in liquid chromatography is supported by a bed either of spherical or irregular particles packed into a stainless steel tube or coated on the internal surface of a quartz capillary tube. Particle diameters (d_p) range from 1 to 37 μm, with 5 μm and 10 μm as the most common sizes used for analytical separations and larger particles ($d_p > 10$ μm) being reserved for preparative separations. Analytical separations are generally conducted in columns whose internal diameter (d_c) ranges from 350 μm to 8 mm. Microbore liquid chromatography is the term usually used for separations conducted with columns of internal diameter equal to or less than 1 mm (Scott, 1980). Preparative liquid chromatography separations are generally conducted in columns of internal diameter greater than 8 mm. Open tubular liquid chromatography is also called capillary liquid chromatography because it is conducted in quartz capillary tubes whose internal diameter ranges from 5 to 50 μm (Ishii, 1988). Single liquid chromatography columns range in length (L) from 5 mm to 1 m. Longer columns can be obtained by linking two or more columns together in series.

2.2 Chromatographic mobility

The moving phase in liquid chromatography is generally referred to as the mobile phase. In liquid chromatography, the mobile phase flows through or across the stationary phase as a result of a pressure difference (P) across the column.

$$P = P_2 - P_1 \qquad (2.1)$$

where P_1 and P_2 are the inlet and outlet pressures, respectively. The solute molecules in liquid chromatography move in the same direction as the solvent molecules at a linear velocity, u, (equation (2.2)) which is equal to or less than the linear velocity of the solvent (u_0) (equation (2.3)):

$$u = \frac{L}{t_{r,i}} \qquad (2.2)$$

$$u_o = \frac{L}{t_o} \qquad (2.3)$$

where t_r and t_0 are the times for solute i and the mobile phase to pass through the column, respectively. Markers injected into the chromatographic system to measure t_0 are referred to as unretained compounds because they do not interact with the stationary phase. Solutes that are too large to enter the pores of the stationary phase support are said to be excluded and will elute from the column before the unretained marker. Any compound that is partially retained by the column will elute after the unretained marker.

Separation of solutes injected into the system arises from differential retention of the solutes by the stationary phase. The net retention of a particular solute depends upon all the solute–solute, solute–mobile phase, solute–stationary phase and stationary phase–mobile phase interactions that contribute to retention. The types of solute–stationary phase interactions involved in chromatographic retention include hydrogen bonding, van der Waal's forces, electrostatic forces or hydrophobic forces.

2.3 Peak shape

2.3.1 Gaussian distribution

Theoretically, a single chromatographic band will assume a Gaussian distribution in which the concentration (C) of the solute is related to the time elapsed from the point of injection of the solute (t) by (Miller, 1988)

$$C = \frac{1}{F_v \sigma} \frac{M}{\sqrt{2\pi}} e^{-\frac{1}{2}\left[\frac{t_r - t}{\sigma}\right]^2} \tag{2.4}$$

where F_v is the volumetric flow rate, σ is the standard deviation of the distribution, M is the mass injected and t_r is the retention time or the time from the point of injection to the peak maximum. Equation (2.4) can be rewritten in terms of σ (equation (2.5)) to show the relationships between the peak width and the standard deviation of the peak,

$$\frac{C}{C_{max}} = e^{\left(\frac{\sigma^2}{2}\right)} \tag{2.5}$$

Figure 2.1 shows that the width of the peak at its base (w_b) is equal to 4σ, which may be obtained by drawing lines at tangents to the points of inflection intersecting the baseline. The points of inflection occur where the width of the peak equals 2σ and $C/C_{max} = 0.607$. The peak width at half height (i.e. $C/C_{max} = 0.5$) is given by

$$w_{0.5} = 2.354 \, \sigma \tag{2.6}$$

2.3.2   Peak area and peak height

Most analytical determinations employing liquid chromatography are based on the measurement of peak area (A) or peak height (h_t). The peak area is equal to the zeroeth moment of the Gaussian distribution (M_o),

$$M_o = A = \int_0^\infty C \, dt \tag{2.7}$$

In practice the peak area (A) is measured by electronic integration of the signal from an on-line detector, in which the response is proportional to the concentration of the solute in the flow cell, according to the relationship

$$A = F \sum_{i=t_{start}}^{i=t_{end}} C_i t_i \tag{2.8}$$

where F is the proportionality factor relating concentration to response. Quantitative determinations may also be based on peak height because the ratio of the peak area to its height is constant (equation (2.9)). Peak heights can be measured electronically or manually from the chart paper.

$$h_t = \frac{A}{\sqrt{2\pi}} \tag{2.9}$$

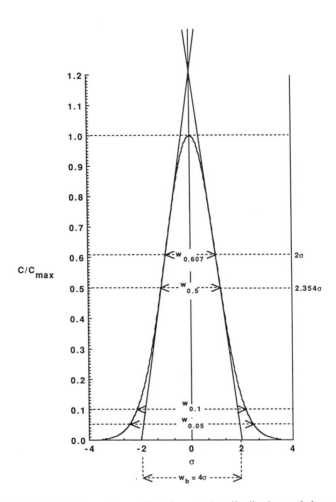

Figure 2.1 Relationships between the widths of a gaussian distribution and the value of σ at various fractions of the peak maximum (C/C_{max}).

2.3.3   Peak asymmetry

Although theory predicts that chromatographic peaks will be symmetrical, peak asymmetry is common, even in the most carefully controlled analytical and preparative separations. The most rigorous definition of peak asymmetry is given by the peak skew (γ), which is related to the second (M_2) and third moments (M_3) of the Gaussian distribution:

$$M_2 = \sigma^2 = \frac{1}{A}\int_0^\infty Ct^2\,\mathrm{d}t - t_r^2 \tag{2.10}$$

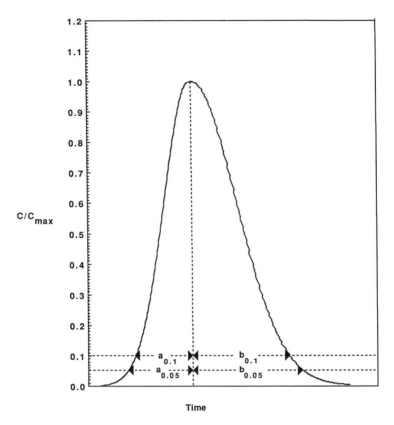

Figure 2.2 Calculation of the front (a) and the tail (b) of an asymmetric peak at 5% and 10% of the peak maximum (equation (2.13)).

$$M_3 = \frac{1}{A} \int_0^\infty C(t^2 - t_r^2)\, dt \tag{2.11}$$

$$\gamma = \frac{M_3}{(M_2)^{0.67}} \tag{2.12}$$

More practical measures of peak asymmetry (A_s) involve the comparison of the width of the tail, b_f, of the peak to its front, a_f (Figure 2.2)

$$A_{s,f} = \frac{b_f}{a_f} \tag{2.13}$$

Recommendations for the position at which $A_{s,f}$ should be measured vary. One of the most rigorous treatments of peak asymmetry is that of Foley and Dorsey (1983) who have described tailing in terms of an exponentially modified Gaussian peak. This group recommend that $A_{s,f}$ be measured at 10% of the peak height, whereas others (including the United States Pharmacopoeia (1990)) recommend that $A_{s,f}$ be measured at 5% of the peal height (see section 2.6.3).

In analytical applications of liquid chromatography the most common causes of peak asymmetry are mixed mechanisms of retention, incompatibility of the sample with the chromatographic mobile phase, or development of excessive void volume at the head of the column. In preparative applications of liquid chromatography and related techniques, column overload can also contribute to peak asymmetry. The causes of severe peak asymmetry in analytical applications should be identified and corrected because they are frequently accompanied by concentration-dependent retention, non-linear calibration curves and poor precision. In addition, peak asymmetry can significantly compromise column efficiency leading, in turn, to reduced resolution and lower peak capacity (see sections 2.5 and 2.6).

2.4 Retention relationships

2.4.1 Retention time (t_r)

The retention time of a chromatographic peak is defined by the first moment of the Gaussian distribution (M_1, equation (2.14)) and is measured from the point of injection to the peak maximum.

$$M_1 = t_r = \frac{1}{A} \int_0^\infty Ct \, dt \qquad (2.14)$$

However, most analytical applications require a definition of retention that is independent of system variables such as column dimensions and flow rate.

2.4.2 Retention volume (V_r)

The retention volume of a peak is equal to the volume of liquid that passes through the column from the point of injection to the point at which the peak maximum exits the column. It is related to the flow rate and the retention time by the equation

$$V_r = F_v t_r \qquad (2.15)$$

2.4.3   Relative retention parameters

Although the retention volume is independent of flow rate, relative retention parameters are preferred because they utilize dimensionless parameters as well as providing additional information about the
chromatographic process. The relative chromatographic mobility (R_c) in
liquid chromatography is defined by

$$R_c = \frac{u}{u_o} = \frac{t_o}{t_r} \tag{2.16}$$

where t_0 is the elution time of an unretained marker. It can be shown that
the relative chromatographic mobility is equal to the fraction of time
spent by the solute molecule in the mobile phase (f_m), which is also equal
to the fraction of the number of molecules in the mobile phase at any
time,

$$R_c = f_m = \frac{N_m}{N_m + N_s} \tag{2.17}$$

2.4.3.1 Capacity factor (k').   The fundamental dimensionless measure of
retention in liquid chromatography is the capacity ratio (or capacity
factor) (k'), which is defined as the ratio of the number of molecules of
solute in the stationary phase, N_s, to the number of molecules in the
mobile phase, N_m

$$k' = \frac{N_s}{N_m} \tag{2.18}$$

By taking into account the volumes of the mobile phase (V_m) and
stationary phase (V_s) in the column, the capacity ratio can be related to
the partition coefficient (K_D) of the solute between the mobile phase and
stationary phase

$$k' = K_D \frac{V_s}{V_m} \tag{2.19}$$

The ratio, V_s/V_m, is often referred to as the phase volume ratio. Combining equations (2.16)–(2.18) allows the development of equations (2.20)–
(2.22), which relate the capacity ratio of a solute to its retention time and
the elution time, t_o, of an unretained compound:

$$k' = \frac{(t_r - t_o)}{t_o} \tag{2.20}$$

$$k' = \frac{t_r}{t_o} - 1 \tag{2.21}$$

$$t_o = t_o(k' + 1) \tag{2.22}$$

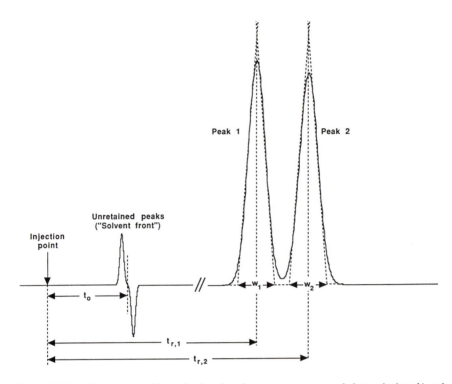

Figure 2.3 Two chromatographic peaks showing the measurements needed to calculate k' and R_s.

Thus defining retention in terms of k' allows calibration of the time scale of a chromatogram in units of t_0 (equation (2.22)) as well as allowing the retention of the solute to be quantified in terms of the extent of its inter-action with the stationary phase (equation (2.19)).

2.4.3.2 Relative retention time (R_t). The main practical disadvantage of using k' to describe retention arises from the difficulties associated with the accurate measurement of t_0. The accurate measurement of t_0 requires correct choice of a reference compound that is neither retained by the column nor excluded from its pores. Problems with the accurate and reproducible measurement of k' have led to the use of a retained com-pound for system suitability tests for liquid chromatography and related techniques when it has been checked that the column, mobile phase and equipment are performing identically to when the method being used was first developed. The relative retention time of a solute j (R_t) may be defined by

$$R_t = \frac{t_{r,j}}{t_{r,i}} \tag{2.23}$$

where $t_{r,j}$ and $t_{r,i}$ are the retention times of the solute j and the reference compound i, respectively. Using retained markers in system suitability tests has the added advantage that such compounds may also be used to check resolution (see section 2.6).

2.5   Band broadening and column efficiency

2.5.1   Number of theoretical plates (N)

The process of band broadening (Figure 2.1) is measured by the column efficiency or the number of theoretical plates (N, equation (2.24)), which is equal to the square of the ratio of the retention time to the standard deviation of the peak. In theory, the value of N for packed columns has only a small dependency on k' and may be considered to be a constant for a particular column. Column efficiency in open-tubular systems decreases markedly with increased retention. For this reason open-tubular liquid chromatography systems must be operated at relatively low k' values (see section 2.5.5.2).

$$N = \left(\frac{t_r}{\sigma}\right)^2 \tag{2.24}$$

where

$$\sigma = \frac{w_{0.5}}{2.354} = \frac{w_b}{4} \tag{2.25}$$

Substitution of σ by either $w_b/4$ or $w_{0.5}/2.354$ (equation (2.25)) gives equations (2.26) and (2.27), respectively. Either equation (2.26) or equation (2.27) may be used for the calculation of column efficiency; however, the latter is generally preferred because it does not require extrapolation of lines at a tangent to the points of inflection (Figure 2.1).

$$N = 16\left(\frac{t_r}{w_b}\right)^2 \tag{2.26}$$

$$N = 5.54\left(\frac{t_r}{w_{0.5}}\right)^2 \tag{2.27}$$

2.5.2   Height equivalent to one theoretical plate (HETP, H)

The number of theoretical plates is directly proportional to the column (L) and inversely proportional to the diameter of the particles (d_p). Comparisons of the quality of two columns of different length may be made by use of the height equivalent to one theoretical plate (HETP, H) which is given by equation (2.28).

$$H = \frac{L}{N} \qquad (2.28)$$

2.5.3   Reduced plate height (h)

Comparisons of columns of different lengths that are also packed with different sized particles may be made by use of the reduced plate height (h) which is a dimensionless parameter defined by equation (2.29).

$$h = \frac{H}{d_{\mathrm{p}}} \qquad (2.29)$$

2.5.4   Relationships between column efficiency, analysis time and back pressure

At first sight one might consider that the most useful liquid chromatography systems are those that employ very high column efficiencies. Increasing the column length or decreasing the particle size represent two potential approaches to the generation of higher column efficiencies (equations (2.27)–(2.29)). Unfortunately, both of these approaches are restricted by the practical limitation of pressure drop across the column (equation (2.1)) which is proportional to the length of the packed bed and inversely proportional to the particle diameter. Most commercial liquid chromatography pumps have an upper pressure limit of about 5000 psi (≈ 340 atm). Furthermore, it is difficult to avoid seal wear and excessive leaks when operating continuously at 5000 psi. Therefore, most methods for routine analysis are developed with pressures of less than 3000 psi (≈ 200 atm). Operating at high pressures for long periods of time can also result in the development of voids in the column, resulting in a loss of column efficiency and deterioration of peak shape.

The pressure drop (P) is related to column length (L) and the linear velocity (u_{o}) by the equation

$$P = \frac{u_{\mathrm{o}} \eta L}{K} \qquad (2.30)$$

where η is the viscosity and K is the permeability of the column. The increased pressure arising from increasing the length of the column or decreasing the particle size can be offset by decreasing the linear velocity, but this can only be done at the expense of longer analysis times. The relationship between column length, linear velocity and analysis time is obtained by combining equations (2.3) and (2.22):

$$t_r = \frac{L}{u_o}(k' + 1) \tag{2.31}$$

A very thorough treatment of the interrelationships between analysis time, pressure drop and column efficiency has been conducted by Knox (1977) and Bristow and Knox (1977) who also considered the complex relationship between linear velocity and column efficiency (section 2.55). Knox (1977) concluded that the best compromise of column efficiency, pressure drop and analysis time is obtained with relatively short columns ($L = 20$–$40\,mm$) that are packed with particles of 1–$2\,\mu m$ in diameter. Such columns can be eluted under gentle conditions (300 psi; 20 atm) and generate 5000–8000 theoretical plates. Similar columns of 30 mm in length packed with $3\,\mu m$ particles have been developed commercially and are sometimes known as 'fast LC' columns because they can be used for rapid analyses. The main disadvantage of these small columns is that relatively minor deterioration of the packed bed can have a significant effect on column performance, thus column life may be short. Modern liquid chromatography columns are typically 10 or 15 cm in length and are packed with $3\,\mu m$ or $5\,\mu m$ particles. Longer columns ($L = 25$–$30\,cm$) packed with $10\,\mu m$ particles will give equivalent separations in terms of column efficiency, but analysis times will be longer (equation (2.31)) and detectability will be compromised.

Knox (1977) also indicated that high column efficiencies can be generated rapidly by using open tubes containing no particles, an idea well established in GC, but unexplored at that time in liquid chromatography. The early 1980s saw the development of microbore liquid chromatography (Scott, 1980), open tubular liquid chromatography (Ishii, 1988) and capillary electrophoresis (Jorgenson and Lukacs, 1981), techniques that allow the generation of more than 100 000 theoretical plates with reasonable analysis times (see section 2.5.5.2). However, open tubular liquid chromatography has not replaced conventional liquid chromatography because it requires specialized instrumentation and is limited to separations with low k' values (section 2.5.5.2). Capillary electrophoresis uses an instrumentional format very similar to that of open tubular liquid chromatography but relies on a different mechanism of separation.

2.5.5   Relationship between mobile-phase velocity and column efficiency

2.5.5.1 Packed columns. Much has been written on the relationships between linear velocity (or flow rate) and column efficiency. The relationship of most practical utility is the Knox equation (1977) (equation (2.32)), which describes the relationship between the reduced plate height (h) and the reduced velocity (v):

$$h = A\,v^{0.33} + \frac{B}{v} + C\,v \tag{2.32}$$

where

$$v = \frac{u\,d_p}{D_m} \tag{2.33}$$

and D_m is the diffusion coefficient of the solute in the mobile phase. The Knox equation, which is analogous to the van Deemter equation used in GC, was developed by taking into account all the possible contributions to band broadening (Figure 2.4). Thus, the A term takes into account the flow inequalities around the particles, the B term considers the longitudinal diffusion of the solute and the C term considers the various mass transport contributions in the stationary phase and stagnant regions of mobile phase. Figure 2.4a shows that the optimum values of A (1), B (2) and C (0.1) in the Knox equation (equation (2.32)) translate to a minimum value of $h \approx 2$ and $v \approx 3$. Table 2.1 shows that for a conventional liquid chromatography column packed with 5 µm particles the optimum flow rate is approximately 0.4 ml/min, which permits an analysis time of about 16 min for $k' = 5$ (see section 2.6). However, operating at a flow rate of 1 ml/min ($n = 7$) will cause only a 20% reduction in column efficiency which will have a relatively small effect on resolution because resolution is proportional to the square root of the number of theoretical plates (see section 2.6).

At most practical flow rates, the relationship between h and the reduced linear velocity is dominated by the A term, which is mainly determined by how well the particles are packed into the column. Figure 2.4b shows how poor column packing and poor mass transfer characteristics will effect column efficiency. Poor column packing ($A = 4$) will have a significant effect on column performance at all flow rates, and this should be suspected if a value of greater than 5 for h is obtained at $v \approx 3$. In contrast to the effects of poor column packing, poor mass transport will result only in low column efficiencies at higher flow rates. However, when testing columns it is very important to minimize extra-column effects, which can contribute significantly to column efficiency, especially at low k' values (see section 2.5.6).

2.5.5.2 Open tubular systems. Using open tubes containing no particles substantially increases the column efficiency because it effectively eliminates the contribution of flow inequalities to band broadening. Thus in open tubular chromatography, the relationship between the reduced plate height and the reduced linear velocity is given by (Golay, 1958):

$$h = \frac{B}{v} + C\,v \tag{2.34}$$

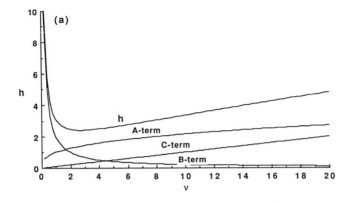

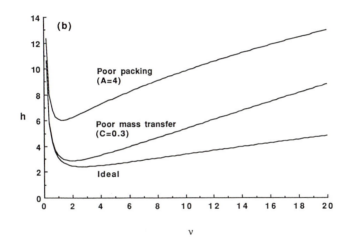

Figure 2.4 Relationship between reduced plate height (h) and reduced velocity (v) according to the Knox equation ($h = Av^{0.33} + B/v + Cv$) (equation 2.32)). (a) Optimum values for the Knox coefficients $A = 1$, $B = 2$, $C = 0.1$; (b) comparison of the ideal curve ($A = 1$, $B = 2$, $C = 0.1$) with that of a poorly packed column ($A = 4$, $B = 2$, $C = 0.1$) and a column with poor mass transfer characteristics ($A = 1$, $B = 2$, $C = 0.3$).

where h and v in open-tubular systems are given by equations (2.35) and (2.36), respectively.

$$h = \frac{H}{d_c} \tag{2.35}$$

$$v = \frac{u\, d_c}{D_m} \tag{2.36}$$

where d_c is the diameter of the column. The value of B in open tubular liquid chromatography is the same as for packed systems because it

Table 2.1 Relationship between flow rate (F_v), linear velocity (u), analysis time and column efficiency in liquid chromatography using a conventional packed column (5 μm, 100 × 5 mm i.d.)

F_v (mL/min)	u (cm/s)	v	t_0 (s)	t_r (min) (for $k' = 5$)	h
8	1.12	56	8.5	0.9	9.41
4	0.56	28	17	1.6	5.87
2	0.28	14	35	3.5	3.95
1	0.14	7	70	7.0	2.89
0.42	0.06	3	163	16.0	2.41
0.2	0.03	1.4	350	35.0	2.70
0.1	0.01	0.7	700	70.0	3.82

depends only on the diffusion coefficient of the solute. However, in contrast to packed-column liquid chromatography, the C term in open tubular liquid chromatography is strongly dependent on k':

$$C = \frac{1}{96} \frac{(1 + 6k' + 11k'^2)}{(1 + k')^2} \qquad (2.37)$$

Equation (2.37) predicts that $h_{min} = 0.28$ at $v = 14$ at $k' = 0$, but h_{min} is increased by a factor of 3–0.84 at $v = 4.7$ at $k' = 5$. Thus, open-tubular systems are limited to separations of relatively low k' values. Further limits are placed on the wider application of open tubular liquid chromatography by its very low sample loadability and the need to minimize extra-column band broadening (section 2.5.6). Nevertheless, some spectacular column efficiencies have been reported using open tubular liquid chromatography. For example, Manz and Simon (1986) have been able to generate 10^6 theoretical plates in approximately 150 s for an unretained solute, KCl. It is worth nothing that the peak width at half height was approximately 600 ms, but that the detector volume needed to avoid extra-column band broadening was 6.2 pl.

2.5.6 Extra-column contributions to band broadening

Relationships between mobile phase velocity and column efficiency, such as the Knox equation (equation (2.32)), are developed by taking into account all the possible contributions to band broadening. Thus the total peak variance (σ^2_{column}) arising from band broadening in a packed column is given by

$$\sigma^2_{column} = \sigma^2_d + \sigma^2_{sm} + \sigma^2_s + \sigma^2_e + \sigma^2_m \qquad (2.38)$$

where the terms on the right-hand side of equation (2.38) correspond to the contributions from longitudinal diffusion (d), stagnant mobile phase mass transfer (sm), stationary phase mass transfer (m), Eddy diffusion (e) and mobile phase mass transfer (m), respectively. The total band broadening (σ^2_{tot}) arises from spreading of the peak in the column as well as in the various instrumental components outside the column:

$$\sigma^2_{total} = \sigma^2_{column} + \sigma^2_{extracolumn} \qquad (2.39)$$

where the extra-column contributions ($\sigma^2_{extracolumn}$) may arise from spreading in the injector (σ^2_{inj}), the detector (σ^2_{det}) and the connecting tubing (σ^2_{tube}):

$$\sigma^2_{extracolumn} = \sigma^2_{inj} + \sigma^2_{det} + \sigma^2_{tube} \qquad (2.40)$$

Because band broadening increases with retention (equation (2.24)) extra-column contributions to band broadening may be more apparent with peaks of low k' values, becoming relatively insignificant for well retained peaks. Reducing extra-column effects to a minimum for conventional liquid chromatography columns (e.g. 5 µm particles, 150 × 5 mm, i.d.) by keeping the total extra-column volumes to less than 10% of the peak volume is relatively straightforward. However, maintaining the extra-column volumes below 10% of the peak volume is not trivial for microbore and open tubular systems, in which d_c is less than 1 mm. Therefore, microbore and open-tubular systems require special, miniaturized instrumentation.

Theoretically, the injection volume (V_{inj}) of a sample should be substantially less than the peak volume. However, in practice very large injection volumes that are well in excess of the peak volume can be made provided the injection solvent is much weaker than the mobile phase used for elution. This is possible because of zone compression, which causes a sample injected in a weak solvent to be compressed at the head of the column. Zone compression can usually be achieved if the concentration of organic modifier in the sample solution is less than 50% of the concentration of the organic modifier in the mobile phase. Even if the sample is dissolved in the mobile phase, substantial zone compression will occur if the chromatographic conditions are arranged such that $k' \geqslant 5$.

2.6   Separation

The separation of two adjacent components in a chromatogram may be described by the selectivity (α) and resolution (R_s) (Figure 2.5a).

2.6.1   Selectivity (α)

The selectivity factor for two adjacent peaks in a chromatogram is defined by

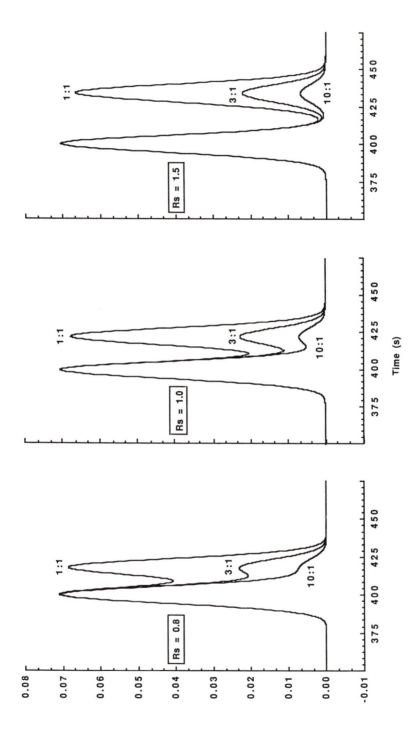

Figure 2.5 Effect of relative concentration on the separation of two components at three values of resolution (R_s). The curves were simulated using equations (2.4), (2.24) and (2.22) and the following parameters: (a) $R_s = 0.8$; $N = 5000$; $k'_1 = 3$; $t_0 = 100$ s; $t_{r,2} = t_{r,1} + 3.2\sigma_1$; (b) $R_s = 1.0$; $N = 5000$; $k'_1 = 3$; $t_0 = 100$ s; $t_{r,2} = t_{r,1} + 4.0\sigma_1$; (c) $R_s = 1.5$; $N = 5000$; $k'_1 = 3$; $t_0 = 100$ s; $t_{r,2} = t_{r,1} + 6.0\sigma_1$.

$$\alpha = \frac{k'_2}{k'_1} \qquad (2.41)$$

where k'_1 and k'_2 are the capacity ratios of the first and second peaks to elute, respectively. Substitution of equation (2.20) into equation (2.41) followed by rearrangement allows the selectivity factor to be expressed in terms of the retention times of the two solutes.

$$\alpha = \frac{t_{r,2} - t_o}{t_{r,1} - t_o} \qquad (2.42)$$

Substitution of equation (2.41) into equation (2.19) gives

$$\alpha = \frac{K_{D,2}}{K_{D,1}} \qquad (2.43)$$

which demonstrates that separation in liquid chromatography arises from differences in partition coefficient (K_D).

2.6.2   Resolution (R_s)

Defining separation in terms of the selectivity factor (equations (2.41)–(2.43)) ignores the effects of peak width and band broadening. The resolution factor (R_s) is defined by equation (2.44), which takes into account both the difference in retention as well as the average peak widths of the two peaks ($0.5(w_{b,1} + w_{a,1})$) (Knox, 1977).

$$R_s = \frac{t_{r,2} - t_{r,1}}{0.5(w_{b,1} + w_{b,2})} \qquad (2.44)$$

Figure 2.5 shows that relative concentration (or for analytical separations chromatographic response) must also be taken into account when considering resolution. In particular it should be noted that the position of the valley between the peaks is shifted to longer times as the relative concentration of the second peak is decreased. When the peak separation is 3.2σ, then $R_s = 0.8$; and no valley is detected if the ratio of the concentrations of the two peaks is 10:1. When the separation of the peak maxima is equal to 6σ, then $R_s = 1.5$, and the degree of peak overlap is less than 1%.

2.6.2.1 Factors influencing resolution.   If the difference in retention times of the two peaks is small ($\leqslant 6\sigma$), then the resolution factor may be approximated by

$$R_s \approx \frac{t_{r,2} - t_{r,1}}{w_{b,1}} \qquad (2.45)$$

By substituting equations (2.26), (2.41), and (2.42) into equation (2.45), equation (2.46) may be obtained

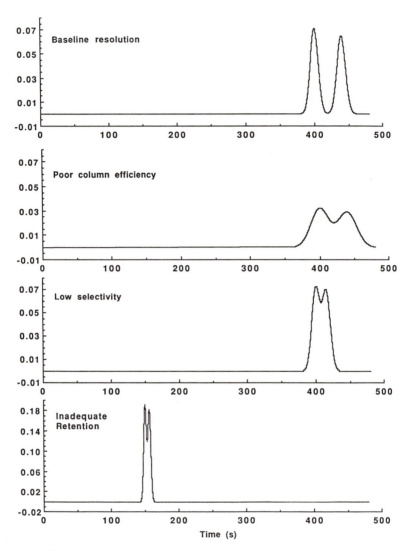

Figure 2.6 Effect of selectivity (α), column efficiency (N) and retention (k') on resolution (adapted from Snyder and Kirkland (1986)). The curves were simulated using equations (2.4), (2.24) and (2.22) and the following parameters: (a) $k'_1 = 3.0$; $t_0 = 100$ s; $\alpha = 1.13$; $N = 5000$; $R_s = 1.53$; (b) $k'_1 = 3.0$; $t_0 = 100$ s; $\alpha = 1.13$; $N = 1000$; $R_s = 0.68$; (a) $k'_1 = 3.0$; $t_0 = 100$ s; $\alpha = 1.05$; $N = 5000$; $R_s = 0.63$; (b) $k'_1 = 0.5$; $t_0 = 100$ s; $\alpha = 1.13$; $N = 1000$; $R_s = 0.67$.

$$R_s = \frac{1}{4} \sqrt{N} \frac{k'_1}{(k'_1 + 1)} (\alpha - 1) \tag{2.46}$$

which allows resolution to be described into terms of a retention factor ($k'/(k' + 1)$), a selectivity factor ($\alpha - 1$) and a column efficiency factor ($N^{0.5}$). If the assumption that the two peak widths are unequal is not made, then a very similar expression (equation (2.47)) is obtained

$$R_s = \frac{1}{4} \sqrt{N} \frac{k'_1}{(k'_1 + 1)} \frac{(\alpha - 1)}{\alpha} \tag{2.47}$$

Of the three factors governing resolution (equations (2.46) and (2.47)), selectivity is the most important and is the one that is most easily manipulated to optimize separations. However, Figure 2.6 shows that poor resolution can arise as a result of poor column efficiency (Figure 2.6b), low selectivity (Figure 2.6c) or inadequate retention (Figure 2.6d). Thus an important part of any optimization strategy is to recognize which parameter is the main contributor to poor resolution.

Figure 2.7 shows that the optimum range for k' is between 2 and 10 and ideally the solvent strength should be adjusted so that the retention of the peak of interest lies within this range. Clearly, decreasing the solvent strength to increase k' to value of greater than 10 has an insignificant effect on resolution. The curves in Figure 2.7 have been drawn using a relatively conservative value of 5000 for N. It is clear that with this modest level of column efficiency baseline resolution ($R_s = 1.5$) of two peaks can be achieved with a selectivity of 1.10 and $k' = 10$. However, if $\alpha = 1.05$ then baseline resolution is not achievable.

Increasing column efficiency (N) is the least attractive method of increasing resolution because it can only be achieved with packed columns at the expense of increased pressure and longer analysis times (see section 2.5.4). Furthermore, equations (2.46) and (2.47) show that resolution is proportional to the square root of the column efficiency. Therefore, doubling the column efficiency by doubling the length or halving the particle diameter only increases resolution by a factor of 1.414. Table 2.2 further illustrates the role of column efficiency in determining resolution showing the number of theoretical plates required for baseline resolution as a function of k' and α. Conventional packed liquid chromatography columns will given column efficiencies in the range 5000 to 15000 and open-tubular systems are generally needed for higher plate counts.

2.6.3 *Effect of peak asymmetry on column efficiency and separation*

Foley and Dorsey (1983) have considered the effect of peak tailing on column efficiency and concluded that measuring peak widths at half height (equation (2.26)) substantially overestimates the number of theoretical plates of a tailed peak. Thus resolution (equations (2.46) and (2.47))

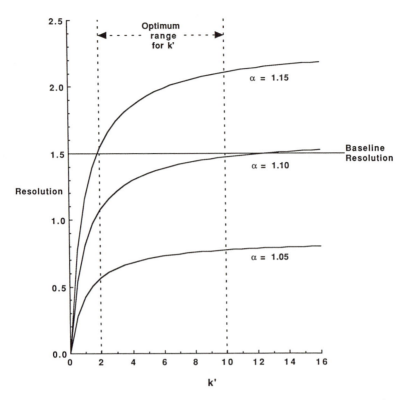

Figure 2.7 Relationship between resolution (R_s), capacity factor (k') and selectivity (α). The curves were simulated using equation (2.47) with $N = 5000$.

Table 2.2   Number of theoretical plates required (N_{req}) to give baseline resolution ($R_s = 1.5$) as a function of the selectivity (α) and retention (k')

Selectivity (α)	N_{req}[a]		
	$k' = 1.0$	$k' = 2.0$	$k' = 5.0$
1.50	1 300	700	500
1.20	5 000	3 000	2 000
1.15	8 500	5 000	3 000
1.10	17 000	10 000	6 000
1.05	63 000	30 000	23 000
1.02	375 000	210 000	135 000
1.01	> 1 000 000	800 000	500 000

$$^a\ N_{req} = \left\{ 6\left(\frac{\alpha}{\alpha-1}\right)\left(\frac{k'+1}{k'}\right) \right\}^2.$$

based on values of N measured by the half-height method will be over-estimated. Accordingly, they recommend that column efficiency of tailed peaks be measured from the peak width at 10% of the maximum ($w_{0.1}$) and that a correction be made for the asymmetry factor measured at the same position

$$N_{sys} = \frac{41.7\left(\dfrac{t_r}{w_{0.1}}\right)^2}{A_{s,0.1} + 1.25} \tag{2.48}$$

Equation (2.48) was derived empirically but for symmetrical peaks ($A_{s,0.1} = 1.00$), equation (2.48) and the corresponding equation obtained assuming a Gaussian distribution differ by only 0.6%:

$$N = 18.42\left(\frac{t_r}{w_{0.1}}\right)^2 \tag{2.49}$$

BIBLIOGRAPHY

Anon (1990) *United States Pharmacopeia. National Formulary*, United States Pharmacopeial Convention, Rockville, MD, pp. 1565–1566.

Bristow, P.A. and Knox, J.H. (1977) Standardization of test conditions for high performance liquid chromatography columns, *Chromatographia* **10**, 279–289.

Foley, J.P. and Dorsey, J.G. (1983) Equations for calculation of chromatographic figures of merit for ideal and skewed peaks, *Anal. Chem.* **55**, 730–737.

Golay, M.J.E. (1958) In *Gas Chromatography 1958*, ed. D.H. Desty, Butterworths, London, p. 36.

Ishii, D. (1988) *Introduction to Microscale High-Performance Liquid Chromatography*, VCH, New York.

Jorgenson, J.W. and Lukacs, K.D. (1981) Zone electrophoresis in open-tubular capillaries, *Anal. Chem.* **53**, 1298–1302.

Knox, J.H. (1977) Practical aspects of LC theory, *J. Chromatogr. Sci.* **15**, 352–364.

Manz, A. and Simon, W. (1986) Potentiometer detector for fast high-performance open-tubular liquid chromatography, *Anal. Chem.* **59**, 74–79.

Miller, J.M. (1988) In *Chromatography: Concepts and Contrasts*, Wiley–Interscience, New York, pp.13.

Scott, R.P.W. (1980) Microbore columns in liquid chromatography, *J. Chromatogr. Sci.* **18**, 49–54.

Snyder, L.R. and Kirkland, J.J. (1979) *Introduction to Modern Liquid Chromatography*, 2nd edn., Wiley–Interscience, New York, pp. 38–39.

3 Modes of chromatography

C.M. RILEY

3.1 Terminology

The terminology of chromatography can be confusing because several terms are often used to describe the same process or system. For example, normal-phase chromatography, straight-phase chromatography and adsorption chromatography are all terms that have been used to describe systems involving a polar stationary phase such as silica gel or alumina and less polar mobile phase comprised of an organic solvent mixture. Sometimes terms are used inappropriately. For example, bonded-phase chromatography and partition chromatography have been used to describe reversed-phase chromatography systems involving hydro-carbonaceous stationary phases and hydro-organic mobile phases. However, covalently bonded phases can also be used in the normal-phase mode, as well as for ion-exchange chromatography. Partition chroma-tography, on the other hand, is a term introduced in the 1950s to describe systems in which the stationary phase is a liquid physically absorbed onto an inert support. In fact, partition chromatography, also called liquid–liquid chromatography, can be operated in the normal-phase or the reversed-phase modes depending on the relative polarities of the mobile phase and the stationary phase. Because it is difficult to maintain the stability of absorbed-liquid stationary phases and the contrasting stability of chemically bonded stationary phases, the use of the true liquid–liquid chromatography systems has largely disappeared and will not be discussed in great detail.

The modes of chromatography that are discussed in this chapter are reversed-phase chromatography, normal-phase chromatography, ion-exchange chromatography and size-exclusion chromatography. with the exception of size-exclusion chromatography (see section 3.8), these modes of chromatography are defined essentially according to the nature of the interactions between the solute and the stationary phase, which may arise from hydrogen bonding, van der Waals forces, electrostatic forces or hydrophobic forces. However, it is important to note that stationary phases are rarely homogeneous and it is not unusual for solutes to be retained by a mixture of interactions. This is especially true for those phases prepared by the covalent attachment of a hydrocarbonaceous func-tional group to a solid support such as silica gel. Mixed mechanisms of interaction with the stationary phase can lead to unpredictable retention behavior. Asymmetrical peaks and concentration-dependent retention times can result from mixed retention mechanisms if one of the sites on the stationary phase is relatively inaccessible to the solute or if it is present in low concentrations.

3.2 Chromatographic format

The various formats available in liquid chromatography are summarized in Figure 3.1. Planar liquid chromatography (PLC) can be conveniently divided into paper chromatography, and thin-layer chromatography (TLC) in which the stationary phase is coated as a thin film on a flat surface made of glass or aluminum. Paper chromatography has been largely replaced by more modern methods but TLC is widely used for qualitative analysis and preparative separations. TLC can also be used for quantitative analysis but column-liquid chromatography is generally

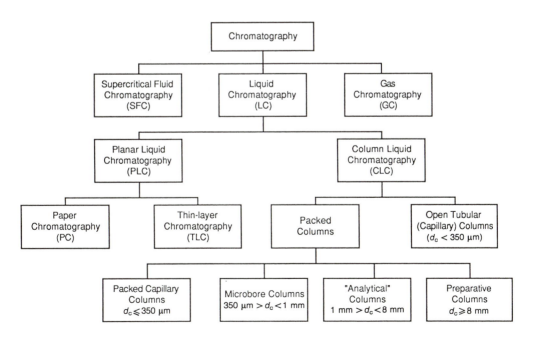

Figure 3.1 The modes of chromatography.

quicker and more efficient. In planar liquid chromatography the mobile phase passes through the stationary phase by capillary action or, less commonly, by centrifugal force.

Column–liquid chromatography (CLC) can be conveniently divided into those systems which use packed columns and those which use open tubes (Figure 3.1). Capillary tubes ($d_c < 350\,\mu m$) are used in open-tubular chromatography and the stationary phase is coated on the internal surface. Packed-column systems can be sub-divided arbitrarily into capillary columns, microbore columns, 'analytical' columns and preparative columns according to the internal diameter of the column (Figure 3.1).

The mobile phase in column liquid chromatography flows through the column as a result of gravitational, mechanical (pumping) or electrical forces. Gravitational or mechanical pressure gradients give rise to a laminar flow profile which is a major cause of band broadening in liquid chromatography (see section 2.5). Electrically driven chromatography is potentially very attractive for the generation of flow in liquid chromatography because it gives rise to a flat flow profile (Figure 3.2). A flat flow profile allows very high column efficiencies to be achieved with short analysis times. The main drawback of electrically driven chromatography is that it has to be conducted in open or packed capillary tubes to minimize the effects of temperature gradients.

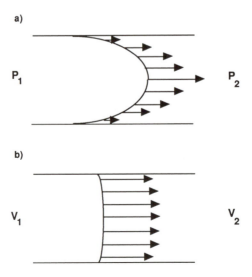

Figure 3.2 Comparison of (a) gravitational or mechanically driven and (b) electrically driven flow profiles.

Electrokinetic chromatography is conducted in capillary tubes and mobility is generated by the application of electrical potential but separation arises as a result of interactions with a micellar 'stationary' phase formed by surfactants added to the mobile phase. Therefore electrokinetic chromatography contains elements of both liquid chromatography and capillary electrophoresis.

3.3   Models of retention

The net retention of an analyte is equal to the sum of the time spent in the mobile phase, t_o, and the time spent in the stationary phase, t_s, i.e.

$$t_r = t_s + t_o \qquad (3.1)$$

The net retention of analyte is a function of all the solute–stationary phase, solute–mobile phase and mobile phase–stationary phase interactions that contribute to retention. It follows that if the interactions with the stationary phase are constant then retention and selectivity are only functions of the composition of the mobile phase. The simplest mobile phase will generally be composed of a mixture of two solvents: a weak solvent A and a strong solvent B. For example a typical reversed-phase eluant will consist of a mixture of water, the weakest solvent in reversed-phase systems, and a stronger organic modifier such as methanol or acetonitrile. Similarly, a normal-phase eluant will typically comprise a mixture of a weak solvent such as a *n*-heptane and a stronger modifier such as chloroform or ethyl acetate.

Two models have been proposed to describe the process of retention in liquid chromatography (Figure 3.3), the solvent-interaction model (Scott and Kucera, 1979) and the solvent-competition model (Snyder, 1968 and 1983). Both these models assume the existence of a monolayer or multiple layers of strong mobile-phase molecules adsorbed onto the surface of the stationary phase. In the solvent-partition model the analyte is partitioned between the mobile phase and the layer of solvent adsorbed onto the stationary-phase surface. In the solvent-competition model, the analyte competes with the strong mobile-phase molecules for active sites on the stationary phase. The two models are essentially equivalent because both assume that interactions between the analyte and the stationary phase remain constant and that retention is determined by the composition of the mobile phase. Furthermore, elutropic series, which rank solvents and mobile-phase modifiers according to their affinities for stationary phases (e.g. Table 3.1), have been developed on the basis of experimental observations, which cannot distinguish the two models of retention.

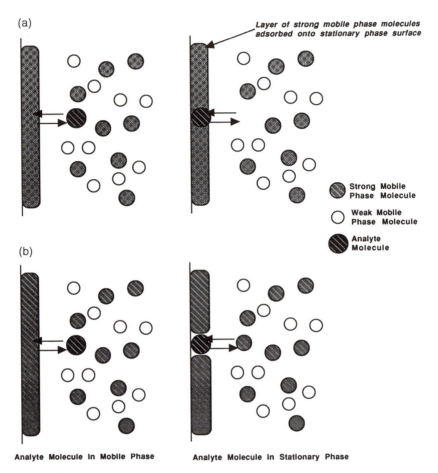

Figure 3.3 Comparison of the (a) solvent-interaction and the (b) solvent-competition models of retention in liquid chromatography.

3.4 Choice of mode of chromatography

Once the chromatographic format for a particular application has been chosen, the strategy for the development of method involves four basic steps: (a) selection of the mode of chromatography, (b) selection of the column type and dimensions (see section 2.5.4), (c) adjustment of solvent strength to achieve a value of k' between 2 and 10 (see section 2.6.2.1) and (d) adjustment of selectivity to achieve $R_s \geqslant 1.50$ (see section 2.6.2.1). Because it is easier to change the composition of the mobile phase than it is to change the nature of the stationary phase, the greatest emphasis in the development of a method should be placed on optimizing the

Table 3.1 Elutropic series for some modifiers commonly used in normal-phase and reversed-phase liquid chromatography, arranged in order of increasing solvent strength

Normal phase		Reversed phase	
Modifier	Solvent strength e^o (silica)[a]	Modifier	Solvent strength $S(C_{18})$[b]
n-Pentane	0.00	Water	–
Carbon tetrachloride	0.18	Methanol	2.6
Diethylether	0.38	Acetonitrile	3.2
Chloroform	0.40	Acetone	3.4
Dichloromethane	0.42	Ethanol	3.6
Tetrahydrofuran	0.45	2-Propanol	4.2
1,2-Dichloroethane	0.49	Tetrahydrofuran	4.5
Acetone	0.56		
1,4-Dioxane	0.56		
Ethyl acetate	0.58		
Dimethyl sulfoxide	0.62		
Acetonitrile	0.65		
2-Propanol	0.82		
Ethanol	0.88		
Methanol	0.95		
Acetic acid	> 1.00		
Water	> 1.00		

[a]Snyder's solvent strength parameter (see Chapter 2, section 2) (Snyder 1968, 1983)
[b]Slope coefficient for equation (3.13) (Poole and Poole, 1991)

composition of the mobile phase. The exception to this is size exclusion chromatography where the separation process is based on differences in molecular size rather than partitioning of the analytes between the mobile phase and the stationary phase.

The mode of chromatography best suited for a given separation is based primarily on the molecular weight, the polarity and the ionic character of the analyte (Figure 3.4). Other factors that should be taken into account when choosing the chromatographic mode include the nature of the sample and whether the separation is to be developed for analytical or preparative purposes; for example, non-polar compounds that are best suited to separation by normal-phase chromatography. However, non-polar compounds are more conveniently analyzed in biological fluids by reversed-phase methods, which use aqueous mobile phases. Another consideration is cost because some preparative columns packed with specialized bonded phases can cost in excess of $10 000. Thus a relatively inexpensive silica column may be preferred for the preparative separation of a very polar compound even if a more suitable analytical method exists using a reversed-phase column.

Traditionally, size-exclusion chromatography (SEC) is recommended for the separation of compounds of molecular weight greater than 1000 and this is still the method of choice for the determination of the size

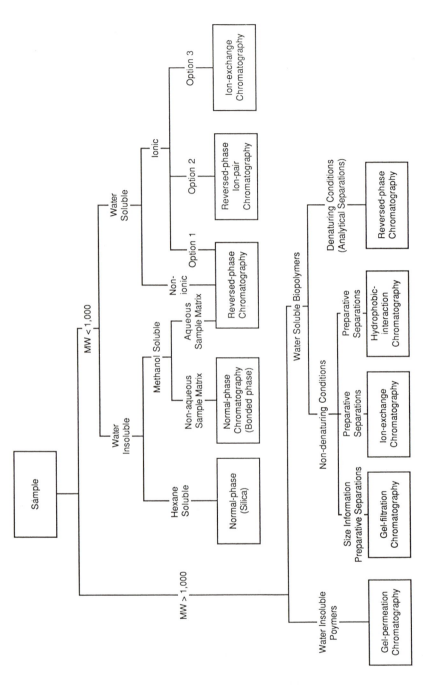

Figure 3.4 Flow diagram used for the choice of chromatographic systems.

distribution of polymers. However, other options exist for the analytical and preparative separations of water-soluble biopolymers such as peptides, proteins or oligonucleotides (Figure 3.4). In addition to size-exclusion chromatography, other non-denaturing systems should be considered for the preparative separations of water-soluble biopolymers including ion-exchange chromatography and hydrophobic-interaction chromatography. Conventional reversed-phase liquid chromatography can also be used for the separation of proteins on stationary phases with large pore diameters; however, the high organic modifier concentrations used in these systems often denature the protein.

The main factors to consider in the choice of mode of chromatography for the separation of small molecules (RMM < 1000) are ionic character and polarity. Ionic compounds are best separated by ion-exchange chromatography or reversed-phase chromatography. In general, reversed-phase chromatography is preferred to ion-exchange chromatography because it usually produces higher column efficiencies and there are more options available for the optimization of separations (see sections 3.6 and 3.7).

The solubility parameter (δ) has been shown to be a useful measure of polarity in liquid chromatography (Wells, 1988; Schoenmakers et al., 1982). For an analyte, i, to elute with a k' value between 2 and 10, its polarity (δ_i) must lie approximately between the polarity of the mobile phase (δ_m) and the polarity of the stationary phase (δ_s). In normal-phase chromatography, $\delta_m < \delta_s$; the value of δ_s for silica gel is 16.0 and the value of δ_m for hexane, the least polar mobile phase, is 7.0. In reversed-phase chromatography the value of δ_s for a typical hydrocarbonaceous support such as octadecyl silica (ODS) is about 7.0 and the value of δ_m for water, the weakest mobile phase, is 23.4. Most applications of liquid chromatography involve the separation of organic molecules whose solubility parameters (δ_i) range between about 8 for non-polar organic compounds such as fatty acids to about 16 for very polar compounds such as 2-hydroxyphenol (Wells, 1988). Thus a wider range of solute polarity can be accommodated by reversed-phase liquid chromatography compared with normal-phase liquid chromatography. This wide range of solute polarity combined with its compatibility with aqueous samples explains why reversed-phase liquid chromatography accounts for the majority of present-day applications (Schoenmakers et al., 1982).

The solubility-parameter approach provides a useful framework for the choice of the mode of liquid chromatography that is best suited for a given analyte. However, solubility parameters of solutes are often unavailable and the solubility of an analyte in water, methanol and hexane provides a more pragmatic basis for the choice of chromatographic system. Figure 3.4 shows that low molecular weight analytes (RMM < 1000) may be divided into those that are water soluble and those that are water insoluble. Water-soluble compounds are further divided into electrolytes and non-electrolytes. On the one hand, non-electrolytes that are also soluble in water are best separated by reversed-phase liquid chromatography. On the other hand, electrolytes can be analyzed by ion-exchange, reversed-phase or reversed-phase ion-pair chromatography. Non-polar compounds that are also soluble in hexane can be analyzed by normal-phase chromatography on silica gel or alumina. More polar compounds that are soluble in methanol but insoluble in water can be separated on a bonded phase by either normal-phase or reversed-phase chromatography.

By virtue of the high polarity of water, reversed-phase chromatography is suitable for the analysis of a much wider range of polarity of compounds than normal-phase chromatography. However, more solvents are available for normal-phase liquid chromatography because the number of suitable organic solvents that are also miscible with water is fairly limited (Table 3.1). Solute-stationary phase interactions in reversed-phase chromatography are relatively non-specific and the number of functionalized stationary phase supports is consequently small. Therefore, the options for the optimization of reversed-phase separations of neutral compounds are somewhat limited. However, one additional advantage of reversed-phase systems not yet mentioned is the opportunity to invoke secondary equilibria and thereby manipulate the retention of ionic compounds. In this context the application of ion-pairing techniques to control the retention of ionic compounds is particularly important.

3.5   Normal-phase liquid chromatography

In normal-phase liquid chromatography the stationary phase is a polar adsorbent and the mobile phase is generally a mixture of non-aqueous solvents.

3.5.1   Retention mechanisms and mobile phase effects in normal-phase liquid chromatography

Both the solvent-interaction model (Scott and Kucera, 1979) and the solvent-competition model (Snyder, 1968, 1983) have been used to describe the effects of mobile-phase composition on retention in normal-phase liquid chromatography. The solvent interaction model on the one hand provides a convenient mathematical model for describing the relationship between retention and mobile phase composition. The solvent competition model on the other hand provides a more complete, quantitative description of the relative strengths of adsorbents and organic solvents used in normal-phase chromatography.

3.5.1.1 Solvent interaction model for normal-phase liquid chromatography. The solvent-interaction model of Scott and co-workers (Scott and Kucera, 1979) assumes that the analyte partitions between the bulk mobile phase and a layer of solvent absorbed onto the stationary phase. The quantitative description of the relationship between retention and the composition of the mobile phase in the solvent-interaction model requires the definition of the void volume corrected retention volume (V'), which is related to the retention volume (V_r) and the void volume (V_0) by

$$V' = V_r - V_0 \qquad (3.2)$$

It will be recalled from equation (2.19) that

$$k' = K_D \frac{V_s}{V_m}$$

If we assume that the volume of stationary phase, V_s, is equal to the volume of adsorbed liquid, V_a, then replacing V_s by V_a and V_m by V_o, and substituting into equation (3.2) gives

$$V' = \frac{K_D V_a V_0}{V_0} = K_D V_a \qquad (3.3)$$

which predicts that V' is proportional to the volume of adsorbed liquid, V_a. If the stationary phase is saturated then retention depends only on interactions in the mobile phase. For binary mobile phases containing solvents A and B the net void volume corrected retention volume of a solute, $V'_{A,B}$ is given by

$$\frac{1}{V'_{A,B}} = \frac{\Phi_A}{V'_A} + \frac{\Phi_B}{V'_B} \qquad (3.4)$$

$$= \Phi_A \left(\frac{1}{V'_A} - \frac{1}{V'_B} \right) + \frac{1}{V'_B} \qquad (3.5)$$

where V'_A and V'_B are the void volume corrected retention volumes of the solute in pure A and B, respectively and Φ_A and Φ_B are the volume fractions of A and B in the mobile phase, respectively. Equation (3.5) can be rewritten in terms of k':

$$\frac{1}{k'_{A,B}} = \Phi_A \left(\frac{1}{k'_A} - \frac{1}{k'_B} \right) + \frac{1}{k'_B} \qquad (3.6)$$

For most polar stationary phases such as silica or alumina eluted with a binary mixture such as chloroform–*n*-heptane, a monolayer of the stronger solvent, in this case chloroform, is formed on the stationary phase over a wide range of Φ values, typically 0.05–0.95. Therefore this model provides a convenient mathematical model relating retention to the composition of the mobile phase. The main disadvantage of this approach is that the values of the constants depend on the nature of the solute as well as the natures of the mobile phase and the stationary phase. The solvent interaction first described by Snyder (1968, 1983) for normal-phase chromatography is more useful for the characterization of the relative strengths of mobile phases and stationary phases.

3.5.1.2 Solvent competition model for normal-phase liquid chromatography. Like the solvent-interaction model, the solvent-competition model assumes that the stationary phase is covered with a monolayer of molecules of the strongest component of the mobile phase. This model also assumes that the concentration of analyte in the stationary phase is small compared with the concentration of solvent molecules and that solute–solvent interactions in the mobile phase are cancelled by identical interactions in the stationary phase. The competition between the analyte molecules, x, and the mobile phase molecules, A, for the active site or sites on the stationary phase is given by

$$x_m + nA_s \rightleftharpoons x_s + nA_m \qquad (3.7)$$

where the subscripts m and s refer to the mobile phase and stationary phases, respectively. The empirical equation relating k' to the solvent strength and to the adsorptive power of the stationary phase was developed by Snyder (equation (3.8)):

$$\log k' = \log\left(\frac{wV_a}{V_s}\right) + \alpha'(S^o - A_i\varepsilon^o) \qquad (3.8)$$

where w is the weight of the adsorbent containing the adsorbed solvent monolayer of volume, V_a, α' is the adsorbent activity parameter ($\alpha' = 1$ for a standard adsorbent), S^o is the dimensionless free energy of adsorption of a standard solute onto a reference adsorbent from a reference solvent and A_i is the molecular cross-sectional area of the analyte, and ε^o is the solvent strength parameter. Both the solvent strength parameter, ε^o, and the solubility parameter, δ_i, are measures of solvent polarity and there is a reasonable degree of correlation between the two. Unlike δ_m, however, the ε^o values take into account specific interactions between solvent molecules and the stationary phase materials. Thus equation (3.8) permits the standardization of both stationary phases and mobile phases. For example, if the k' of a particular test compound is measured in two mobile phase systems, then the relative solvent strength of the two solvent systems, a and b, may be determined from the relationship

$$\log\left(\frac{k'_a}{k'_b}\right) = A_i(\varepsilon^o_b - \varepsilon^o_a) \qquad (3.9)$$

Similarly, the relative activity of two adsorbents, x and y, may be measured with a standard reference analyte and mobile phase system from the relationship

$$\log\left(\frac{k'_x}{k'_y}\right) = S^\circ(\alpha'_x - \alpha'_y) \tag{3.10}$$

The solvent strength parameter for a binary mobile phase, $\varepsilon^0{}_{A,B}$, may be calculated from the equation (Snyder, 1968)

$$\varepsilon_{A,B} = \varepsilon_A + \log\left(\frac{N_B 10^{\alpha n_B}(\varepsilon_B - \varepsilon_A) + 1 - n_B}{\alpha' n_B}\right) \tag{3.11}$$

where N_B is the mole fraction of the strong solvent B and n_B is the adsorption cross-sectional area of solvent B. The main drawback of this approach is that the calculation of solvent strength values for mixed solvent systems is complex (equation (3.11)) and it is generally necessary to refer to tables (e.g. Table 3.2) or monograms for the determination of ε° values of binary mobile phases (Fig. 3.5).

3.5.2 Mobile phase optimization in normal-phase liquid chromatography

The first step in mobile phase optimization involves adjustment of the solvent strength to achieve a k' value between 2 and 10 for all the analytes in the sample to ensure maximum resolution (see section 2.6.2.1). Ideally, components in the sample that are not of interest should also elute with a k' value of less than 10 to minimize the analysis time. A binary solvent system should be employed and the nature of the solvents A and B is relatively unimportant for the establishment of the initial conditions. If the retention times are too long then the mobile phase must be strengthened by increasing the concentration of the stronger solvent B. Conversely, if retention is weak the concentration of solvent B should be increased. Generally, changing the value of ε° by 0.05 will change the k' value by a factor of between 2 and 4. The situation where it is not possi-

Table 3.2 Series of isoelutropic mobile phases for normal-phase liquid chromatography

Solvent	ε^{0a}	Composition (by volume) to give $\varepsilon^\circ = 0.3^c$									
		A	B	C	D	E	F	G	H	I	J
n-Hexane	0.01			25		35		55		65	70
Carbon tetrachloride	0.18	45	55		75		80		85		
Chloroform	0.40	55		75							
Tetrahydrofuran	0.45		45			65					
Acetonitrile	0.65				25			45			
2-Propanol	0.82					20				35	
Methanol	0.95								15	35	30
$\Delta\varepsilon^{0b}$		0.22	0.27	0.29	0.37	0.44	0.64	0.64	0.77		0.94

[a]Solvent strength parameter.
[b]Defined as $\Delta\varepsilon^\circ = \varepsilon^\circ{}_B - \varepsilon^\circ{}_A$.
[c]Adapted from Wells (1988).

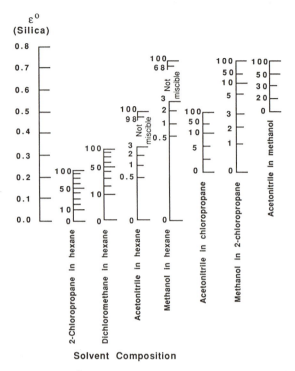

Figure 3.5 Solvent strengths (ε^0 values) various solvent mixtures in normal-phase liquid chromatography.

ble to achieve a k' value of between 2 and 10 under isocratic conditions is known as the general elution problem, which can only be solved by gradient elution or multi-dimensional (column-switching) systems.

Once the correct solvent strength has been identified, fine tuning of the system may be necessary to achieve baseline resolution ($R_s \geqslant 1.5$) between each adjacent pair of peaks in the chromatogram (see section 2.6.2.1). This is usually the most time-consuming phase of method development and is the aspect of computer-assisted optimization methods that has received the most attention to date. The solvent-interaction model of Snyder (equation (3.8)) assumes that solute–solvent interactions in the mobile phase are balanced by equivalent solute–solvent interactions in the mobile phase. Accordingly, isoelutropic mobile phases having the same value of ε^0 should theoretically give the same k' value, independent of the nature of the solvent. Because the activity of the strong solvent in the mobile phase is different from its activity in the stationary phase, the solute–solvent interactions in the stationary phase and mobile phase do not exactly cancel. Consequently, normal-phase liquid chromatographic separations can be optimized by taking advantage of specific solute–solvent inter-

Table 3.3 Physical properties of solvents commonly used in liquid chromatography

Solvent	UV cut-off (nm)	Refrative index (20°C)	Boiling point (°C)	Viscocity (cP,20°C)	Density (20°C)
n-Hexane	190	1.375	68.7	0.31	0.659
Carbon tetrachloride	265	1.460	76.8	0.97	1.594
2-Chloropropane	230	1.378	34.8	0.33	0.862
Chloroform	245	1.446	61.2	0.57	1.480
Dichloromethane	233	1.424	39.8	0.44	1.335
Tetrahydrofuran	212	1.403	66.0	0.55	0.889
Diethyl ether	218	1.352	34.6	0.23	0.713
Ethyl acetate	256	1.372	77.1	0.45	0.901
Dioxane	215	1.422	101.3	1.32	1.034
Acetonitrile	190	1.344	81.6	0.37	0.782
2-Propanol	190	1.378	82.4	2.30	0.785
Methanol	190	1.329	64.7	0.60	0.791
Water	190	1.333	100.0	1.00	0.998
Acetic Acid		1.372	117.9	1.22	1.049

actions. The solvents commonly used in normal-phase liquid chromatography can be divided into selectivity groups according to their hydrogen bonding properties, dipole moment and intrinsic dispersion. Table 3.2 describes the composition of nine binary and one ternary solvent systems, which are isoelutropic ($\varepsilon^o = 0.3$) and can be used as a rough guide to method development in normal-phase liquid chromatography.

It has been shown that the maximum discriminating power of a binary solvent system is achieved if the value of $\Delta\varepsilon^o$ for the two solvents is large. However, the use of binary–solvent mixtures in which $\Delta\varepsilon^o$ is greater than 0.5 can produce poor peak shapes. This is particularly likely if one of the components has a very low ε^o value (e.g. n-heptane) and the amount of residual moisture is not carefully controlled (see section 3.5.3). Lack of miscibility also precludes certain combinations of solvents. Other factors (Table 3.3) that should be taken into account in the choice of mobile phase include viscosity (which will determine the pressure drop (equation (2.30)), boiling point (which will determine the usable range of elution temperatures), and compatibility with the detection system. The high UV cut-off (Table 3.3) of acetone and aromatic hydrocarbons such as benzene and toluene effectively eliminates these and many other solvents from the list of solvents that can be used in column liquid chromatography with UV detection. The UV cut-off of halogenated hydrocarbons ranges from 230 to 265 nm (Table 3.3), and this prohibits their use with UV detection at shorter wavelengths. It should also be noted that halogenated hydrocarbons also quench fluorescence drastically.

3.5.3 Stationary phase effects in normal-phase liquid chromatography

In normal-phase liquid chromatography the stationary phase is typically a polar solid adsorbent eluted with a non-aqueous solvent system. Stationary phases commonly used in normal-phase chromatography include solid adsorbents such as silica gel, alumina and carbon as well as silica gel covalently modified with cyanopropyl-, aminopropyl- and diol functional groups. Separations have also been described on these stationary phases following elution with aqueous mobile phases. However, the retention mechanism in these systems is generally considered to be the same as for reversed-phase systems. The cyanopropyl- and aminopropyl-bonded stationary phases are particularly versatile, being used widely in both the reversed-phase and normal-phase modes. Aminopropyl-bonded phases can also be used as weak anion exchangers (WAX) (see section 3.7).

Silica is generally preferred for normal-phase chromatography because it is more readily available and there is a great deal more information on its chromatographic properties. The adsorbent strength of silica is similar to that of alumina; however, the values of the solvent strength parameters (ε°) for silica are lower by a factor of 0.8 than the corresponding values for alumina. Therefore, the selectivity of silica-based systems is less sensitive to changes in solvent composition than alumina-based systems. Furthermore silica gel is acidic and organic amines are very strongly retained, often eluting as very asymmetric peaks from silica columns. In contrast, amines can be eluted as symmetrical peaks from alumina, which is a basic adsorbent. Similarly, acidic compounds such as carboxylic acids and phenols are well behaved on silica columns but are highly retained on alumina. Although there is generally a good correspondence between the ε° values on silica and alumina, there are some significant differences in the solvent strength parameters, particularly at higher values of ε°, which indicates that subtle differences in selectivity also exist between the two phases.

The solvent-strength parameters for the common solvents used in normal phase chromatography on carbon are quite different from those of silica or alumina. Thus carbon offers quite different selectivities than alumina and silica for normal-phase chromatography. However, the lack of a reproducible commercial source for carbon was for many years a significant limitation to its widespread application. In addition the sensitivity of carbon to changes in solvent strength is much less than that of silica or alumina.

Silica is a rigid polymer consisting of tetrahedral silicon atoms, each attached to four oxygen atoms. The two functional groups of importance on the surface of silica are the Si—O—Si groups (siloxane) and the Si–OH (silanol) groups. In a non-aqueous environment the non-polar siloxane groups are relatively unimportant and the main interactions with the silica support arise from hydrogen bonding and van der Waal's interactions between the solute molecules and the surface silanol groups. In the aqueous environments of typical reversed-phase systems the silanol groups are generally deactivated due to solvation by water molecules. However, interactions with the non-polar siloxane groups can result in retention on silica gel eluted with an aqueous mobile phase. Furthermore, the silanol group is weakly acidic with an apparent pK_a ranging from 4 to 6 depending on its microscopic environment. Thus, the silanol group can behave as a weak ion-exchanger in aqueous (reversed-phase) systems. The ion-exchange properties of residual silanols and, to a lesser extent, their ability to form hydrogen bonds, are responsible for the complex mixed mechanism of retention commonly seen with covalently modified silicas used in reversed-phase liquid chromatography (see section 3.7).

Chemically modified silicas, such as the aminopropyl-, cyanopropyl- and diol phases are useful alternatives to silica gel as stationary phases in normal-phase chromatography. The aminopropyl- and cyanopropyl-phases provide opportunities for specific interactions between the analyte and the stationary phases and thus offer additional options for the optimizations of separations. Other advantages of bonded phases lie in the increased homogeneity of the stationary-phase surface. Silica gel is a heterogeneous material presenting surface silanol groups with a range of activity. Thus coating the surface of the silica gel deactivates the surface producing a more homogeneous surface for normal-phase liquid chromatography. The decreased activity of the cyanopropyl-, aminopropyl- and diol phases compared with silica gel make them more suitable for the separation of moderately polar compounds. However, moderately polar compounds are also well suited to separation by reversed-phase chromatography; therefore, the choice between normal-phase and reversed-phase will often be based on the nature of the matrix rather than properties of the analyte (Figure 3.4).

3.5.4   The role of water in normal-phase liquid chromatography

Water is the strongest solvent in normal phase liquid chromatography (Table 3.1) and the presence of trace amounts of water in an otherwise weak mobile-phase system can have a significant effect on retention in normal-phase liquid chromatography. In fact changes in residual moisture and atmospheric humidity are the most common cause of poor reproducibility of retention time in normal-phase systems. Resolution of this problem is most conveniently achieved by drying the solvents and then adding a constant concentration of water or some very polar modifier such as acetic acid to the mobile phase. The addition of such polar modifiers serves to deactivate the more polar adsorption sites on the surface of the stationary phase, which in turn will improve peak shape as well as the reproducibility of retention times. The same effect can also be achieved by the addition of trace amounts of an amine such as triethylamine (TEA), which is essential for the normal-phase liquid chromatography of amines on silica gel.

3.6 Reversed-phase liquid chromatography

Reversed-phase liquid chromatography is the most popular mode of chromatography for the analytical and preparative separations of compounds of interest in the chemical, biological, pharmaceutical and biomedical sciences (Horváth *et al.*, 1976; Horváth and Melander, 1977; Karger *et al.*, 1976; Hennion *et al.*, 1978; Karger and Geise, 1978; Tanaka *et al.*, 1978; McCormick and Karger, 1980; MacBlane *et al.*, 1987; Schomburg, 1988). The popularity of reversed-phase liquid chromatography derives, in part from the fact that it requires aqueous mobile phases. Therefore, it is generally compatible with most aqueous samples, which can often be injected directly onto the column without pretreatment. If pretreatment of the sample is required, it is often relatively simple. For example the precipitation of proteins and injection of the supernatant is a common preparation procedure for the analysis of blood plasma samples by reversed-phase liquid chromatography. However, the compatibility of the mobile phase with aqueous samples is not the only reason for the widespread acceptability of reversed-phase liquid chromatography. Of equal importance is the wide range of polarity of analytes that can be separated by this technique (section 3.4).

3.6.1 Retention mechanisms in reversed-phase liquid chromatography

The term reversed-phase liquid chromatography derived from the fact that the mobile phase is more polar than the stationary phase which is the opposite of normal-phase chromatography. The reversal of the polarities of the mobile phase and the stationary phase compared with normal-phase chromatography also reverses the order of elution of analytes. A typical reversed-phase system employs a hydrocarbonaceous phase covalently bonded to silica gel and a hydro-organic mobile phase such as a mixture of water and methanol. In addition to silica-based stationary phases, a range of polymeric supports such as polystyrene-divinylbenzene copolymer have been developed for reversed-phase liquid chromatography. The main advantage of the polymeric supports are the lack of residual silanol groups and the greater chemical stability in acidic and basic solutions compared with silica gel. The polymeric supports are chemically stable over the pH range of 1–13 compared with a pH range of 2.5–7.5 for silica-based supports.

3.6.2 Mobile phase effects in reversed-phase liquid chromatography

The main driving force for retention in reversed-phase liquid chromatography is the hydrophobic effect (Horváth *et al.*, 1976; Karger *et al.*, 1976; Horváth and Melander, 1977; Hennion *et al.*, 1978; Karger and

Table 3.4 Hydrophobicities and solubility parameters of bonded stationary phases used in reversed-phase liquid chromatography

Functional group	Log P^a	δ^b
Octadecyl (C$_{18}$)	10.2	8.0
Octyl (C$_8$)	4.81	7.6
Phenyl	3.91	8.6
Hexyl (C$_6$)	3.72	7.3
Propyl (C$_3$)	2.08	6.4
Methyl (C$_1$)	0.99	5.4
Aminopropyl	0.71	8.9
Cyanopropyl	0.39	10.8

[a]Calculated from the 1-octanol/water partition coefficient of the corresponding alkane (Wells, 1988).
[b]Solubility parameter (Wells, 1988; Schoenmakers et al, 1982).

Geise, 1978; Tanaka et al., 1978; McCormick and Karger, 1980; Schoenmakers et al., 1982). Thus retention by the stationary phase arises from repulsion of non-polar (hydrophobic) regions of the solute molecules by the water molecules in the mobile phase. These hydrophobic interactions are modulated by specific solute–solvent interactions in the mobile phase and the stationary phase. As in normal-phase liquid chromatography, retention in reversed-phase liquid chromatography is governed by interactions in the mobile phase because the stationary phase surface is saturated by molecules of the organic modifier added to the mobile phase. Thus the composition of the stationary phase will remain fairly constant with changes in composition of the mobile phase. A monolayer of methanol will adsorb onto the surface of the stationary phase if the concentration of methanol in the mobile phase is greater than 10%. For stronger modifiers such as acetonitrile or tetrahydrofuran, the surface of the stationary phase will be saturated at mobile phase concentrations of less than 10%.

Horváth and co-workers have applied (Horváth et al., 1976; Horváth and Melander, 1977) Sinanoglu's solvophobic theory (Sinanoglu, 1967) to reversed-phase liquid chromatography and that approach probably represents the most rigorous treatment of the subject to date. The application of solvophobic theory to reversed-phase liquid chromatography retains elements of both the solvent competition model and the solvent interactions model, taking into account all the possible solute–solvent–stationary phase interactions contributing to retention. A more detailed description of retention in reversed-phase liquid chromatography is given in papers by Horváth and co-workers (1976 and 1977) on the subject. The discussion presented here involves a more empirical description of the effects of solvent strength on retention and selectivity.

Retention in reversed-phase liquid chromatography decreases with decreasing concentration of water in the mobile phase. The logarithm of the capacity ratio for a given solute is linearly related to the volume fraction of the organic modifier in the mobile phase, Φ, according to equation (3.12):

$$\log k' = \log k'_w - S\Phi \tag{3.12}$$

where k'_w is the capacity ratio of the solute with a completely aqueous mobile phase (i.e. $\Phi = 0$) and S is the slope coefficient. Equation (3.12) provides a convenient framework for the characterization of solvent strength and stationary phase polarity (see section 3.6.2). The slope coefficient, S, forms the basis of an index of solvent strength because it may be related to $\log k'_s$ by

$$S = \log k'_w - \log k'_s \tag{3.13}$$

where k_s is the capacity ratio of a reference solute when $\Phi = 1$. Typical values of S range from 2.6 for methanol to 4.5 tetrahydrofuran (THF) (Table 3.1). Table 3.1 also shows that the requirement that the solvent be miscible with water significantly limits the number of organic modifiers available for reversed-phase liquid chromatography. In fact this list is further reduced by the prohibitive cost of ethanol, the high UV absorbance of acetone and the high viscosity of 2-propanol. Dioxane is rarely used probably because its solvent strength is similar to that of acetonitrile. Therefore, the organic modifiers commonly used for reversed-phase liquid chromatography are, in order of solvent strength: methanol < acetonitrile < tetrahydrofuran (Figure 3.6).

Just as the value of S has been used to define solvent strength the intercept coefficient, $\log k_w$ has been used as a measure of solute hydrophobicity. The use of simple, empirical relationship, such as equation (3.12), for the characterization of chromatographic systems and for the determination of solute hydrophobicity is very attractive. Unfortunately this approach is subject to errors, which arise from the fact that equation (3.12) holds only for relative small ranges of Φ (approximately 0.3). Thus substantial errors in the estimated value of $\log k_w$ can occur if large extrapolations to $\Phi = 0$ are made. Similarly, errors in the relative values of S for different organic modifiers may also arise if they are measured over different ranges of Φ. For ranges of Φ greater than 0.3, $\log k'$ is more appropriately related to Φ by the quadratic relationship (Schoenmakers, 1982)

$$\log k' = \log k'_w + A\Phi + B\Phi^2 \tag{3.14}$$

Restricting the list of organic modifiers to three solvents (methanol, acetonitrile and THF) substantially simplifies method development and the desired strength can generally be achieved with a binary mixture of

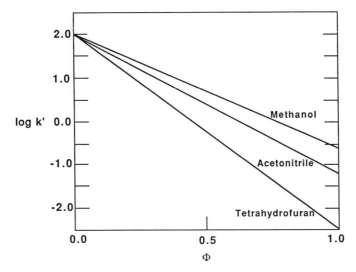

Figure 3.6 Theoretical relationships between the retention ($\log k'$) and the concentrations of methanol, acetonitrile and tetrahydrofuran in mobile phase drawn according to $\log k' = \log k'_w - S\Phi$ (equation (3.12)) using the values of S in Table 3.1.

water and any one of the three organic modifiers. However, having only three organic modifiers limits the options available for the manipulation of selectivity.

Because retention is mainly related to solute hydrophobicity, the 1-octanol/water partition coefficient ($\log P$) has been proposed as a useful predictor of retention in reversed-phase liquid chromatography. Considering first the nature of the solute, there is a generally good correlation between $\log P$ and the $\log k'$ values for solutes on hydrocarbonaceous stationary phases eluted with simple water–methanol mobile phases. The correlation between $\log k'$ and $\log P$ has been used with great success for the prediction of solute hydrophobicity (equation (3.14)) (Karger *et al.*, 1976)

$$\log P = C \log k' + D \qquad (3.14)$$

where C and D are the slope and intercept coefficients, respectively. It will be recalled (see section 2.4.3.1) that the capacity ratio is related to the partition coefficient of the solute between the mobile phase and the stationary phase by

$$k' = K_D \frac{V_s}{V_m}$$

Taking logarithms of both sides of this equation gives

$$\log k' = \log K_D + \log\left(\frac{V_s}{V_m}\right)$$ (3.15)

and substituting equation (3.15) into equation (3.14) gives

$$\log P = C \log K_D + \left[D + \log\left(\frac{V_s}{V_m}\right)\right]$$ (3.16)

which describes the empirical relationship between the chromatographic partition coefficient and the 1-octanol/water partition coefficient.

3.6.2.1 Mobile phase optimization in reversed-phase liquid chromatography. Although the relationship between $\log k'$ and $\log P$ has been used successfully to predict solute hydrophobicity, it has proved less useful for the prediction of optimal chromatographic conditions because it requires prior knowledge of the partition coefficients of the solutes as well as the values of the regression coefficients C and D, which themselves depend on the nature of the mobile phase and the stationary phase. Nevertheless, the $\log P$ approach does provide a useful framework within which to discuss mobile phase optimization strategies in reversed-phase systems. Whereas the correlation between $\log k'$ and $\log P$ is high for systems using methanol–water mobile phases, the correlation for systems using acetonitrile and water–tetrahydrofuran is much weaker. These differences arise from specific solute–solvent interactions, which arise, in turn from the different proton acceptor properties, proton donor properties and dipoles of the three solvents and provide the basis for mobile phase optimization procedures in reversed-phase systems.

As in normal phase (see section 3.5.3), the first step in mobile-phase optimization is the determination of the solvent strength that will elute the analytes with a k' value between 2 and 10 from the chosen stationary phase. It is not important which modifier is chosen to determine the initial conditions, and methanol–water (50:50, v/v) is a convenient starting place. Once the initial conditions have been established, a variety of techniques may be employed to obtain the optimum separation. Most optimization strategies involve the establishment of the isoelutropic concentrations of methanol–water, acetonitrile–water and tetrahydrofuran–water. The isoelutropic concentrations can be determined by experiment or from tables of isoelutropic mixtures (e.g. Table 3.5) (Wells, 1988). The binary solvent systems A, B, C (Table 3.5, Figure 3.7) define the isoelutropic plane, which is then explored to obtain the optimum combination of water, methanol, tetrahydrofuran, and acetonitrile required for the separation.

Assay ruggedness should also be considered in addition to analysis time and resolution in the development of a method. Table 3.1 and Figure 3.7 show that k' is much more sensitive to small changes in the concentration of tetrahydrofuran than to changes in the concentration of methanol or

Table 3.5 Series of isoelutropic mobile phases for reversed-phase liquid chromatography

Solvent[a]	A	B	C	D	E
Methanol	60				
Acetonitrile		46		20	36
Tetrahydrofuran			37	14	8
Water	40	54	63	57	56

[a]Adapted from Wells, (1988); see also Figure 3.7.

acetonitrile. Thus, if the optimization studies reveal several alternatives for mobile phase composition, then the mobile phase with the lowest concentration of tetrahydrofuran and the highest concentration of methanol should be used, because that combination will minimize the variability in retention time arising from batch-to-batch variability in mobile phase composition.

3.6.3 Stationary phase effects in reversed-phase liquid chromatography

The majority of reversed-phase methods have been developed on covalently modified silica gel and the most popular stationary phase is octadecylsilyl silica (ODS, C_{18}). Polymeric supports, such as functionalized polystyrene-divinylbenzene copolymers (MacBlane *et al.*, 1987), are particularly useful when mobile phases of higher pH are required because of their resistance to degradation in alkaline solutions. The main drawback of polymeric supports is their reduced column efficiencies and their lower mechanical resistance to high pressures compared with silica gel.

Because solvophobic interactions dominate retention in reversed-phases systems (Horváth *et al.*, 1976; Karger *et al.*, 1976; Horváth and Melander,

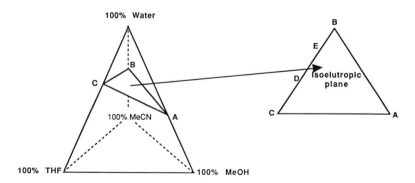

Figure 3.7 Phase diagram for methanol–acetonitrile–tetrahydrofuran–water solvents systems in reversed-phase liquid chromatography, showing the isoelutropic plane, A–B–C, and the positions of the isoelutropic mobile phases D and E described in Table 3.5.

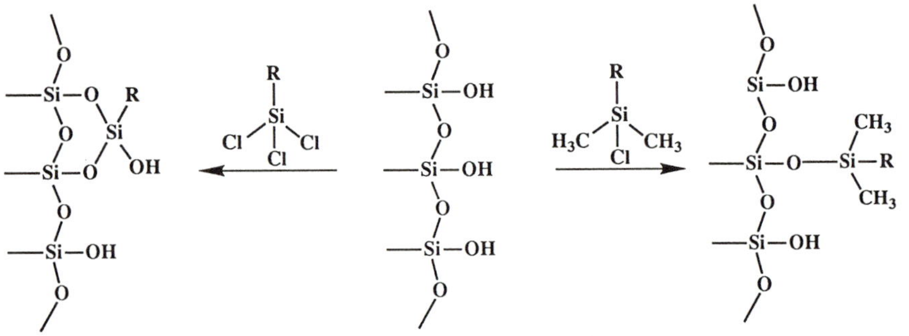

Polymeric

Octadecylsilyl silica (ODS or C_{18}) R=-$C_{18}H_{37}$ Monomeric

Octylsilyl silica (C_8) R=-C_8H_{17}

Phenylsilyl silica (Phenyl) R=-C_3H_6-Ph

Propylsilyl silica (C_3) R=-C_3H_7

(Tri)methylsilyl silica (C_1) R=CH$_3$

1-Aminopropylsilyl silica (Aminopropyl) R=-C_3H_6-NH$_2$

1-Cyanonpropylsillyl silica (Cyanopropyl) R=-C_3H_6-CN

Figure 3.8 Reactions used for the synthesis of polymeric or monomeric (or 'brush' type) bonded stationary phases.

1977), retention on hydrocarbonaceous supports is determined mainly by the carbon loading of the stationary phase (Horváth *et al.*, 1976; Karger *et al.*, 1976; Horváth and Melander, 1977; Hennion *et al.*, 1978; Tanaka *et al.*, 1978). In addition to solvophobic interactions, the retention of certain analytes may also be modulated by specific solute–stationary phase interactions such as dipole–dipole interactions on cyano-bonded phases and π–π stacking interactions on phenyl-bonded phases.

The carbon loading of a stationary phase depends upon the relative surface coverage and the chain length of the bonded functional groups. The surface concentration of silanol groups is between 7 and $8\,\mu\text{mol/m}^2$ of which 3–$4\,\mu\text{mol/m}^2$ are available for covalent attachment of the stationary phase (Majors, 1980). Consequently, the carbon loading is also a function of the surface area of the silica support, which may range between 100 and $400\,\text{m}^2/\text{g}$. The carbon loading will also depend on the method of synthesis. A variety of reactions have been developed for the covalent attachment of hydrocarbons to silica gel. The two reactions most commonly exploited for the attachment of a hydrocarbonaceous phase to silica use either a chlorodimethylalkylsilane or a trichloro-alkylsilane (Figure 3.8). The former produces a monomeric or so-called 'brush-type' stationary phase. The highest carbon loading that can be achieved with a brush-type phase is about 15%. Carbon loadings of greater than 20% can be achieved by the reaction of silica with trichloro-alkylsilane, which produces a polymeric hydrocarbonaceous stationary phase. Polymerization of the stationary phase may have the advantage of protecting the surface of the silica gel from hydrolysis; however, Figure 3.8 shows that both methods of preparing stationary phase result in residual silanol groups at the base of the hydrocarbonaceous functional groups. Interactions with residual silanol groups can contribute to a mixed mechanism of retention. These interactions can occur as a result of electrostatic interactions between cationic solutes and the ionized form of the silanol group or as a result of hydrogen bonding with the neutral form of the silanol. The former type of silanol interaction is important because it can result in severe peak tailing and is discussed in more detail in section 3.6.3.

Theoretically the highest selectivity is achieved in reversed-phase systems when the polarity difference between the mobile phase and the stationary phase is the greatest. In practice, the solvent strength of the mobile phase must be matched with the hydrophobicity of the stationary phase for a given solute to elute with a k' value between 2 and 10. Fortunately, different isoelutropic combinations of mobile phase and stationary phase can give rise to different selectivities because of changes in specific solute–solvent interactions. Changes in selectivity produced by different stationary phases are generally more profound if the differences in structures of the solutes to be separated are associated with polar functional groups (e.g. —OH, —C＝O, —COOH, —NR$_2$, —NO$_2$ etc.). The selectivity of compounds having different non-polar (hydrophobic) functional groups (e.g. —CH$_3$, —CH$_2$—, —Cl, —Br etc.) are much less sensitive to changes in mobile phase and stationary phase under isoelutropic conditions (Karger et al., 1976).

It follows that the polarity difference required to elute a given solute in the ideal k' range can be achieved with various combinations of stationary phase and mobile phase and there is no unique solution for the reversed-phase liquid chromatography of many analytes of intermediate polarity. The differences in surface coverage, bonding chemistry and residual silanols help to explain the dramatic differences that can exist between the same type of stationary phase produced by different manufacturers. Furthermore, a given manufacturer may produce several types of stationary phase containing the same functional group attached to the silica gel. In the early days of bonded phase chromatography, batch-to-batch variability was a significant problem. However, advances in bonding chemistry have significantly improved the batch-to-batch reproducibility of bonded stationary phases.

Exploiting the differences in the characteristics of various commercially available stationary phases would appear to be an attractive option for the development of a reversed-phase method. However, most optimization

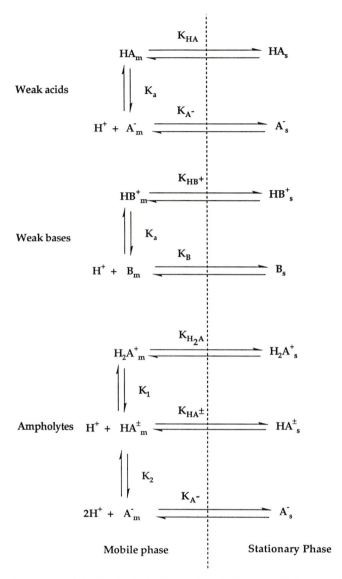

Figure 3.9 Chromatographic distribution of a weak acid in reversed-phase liquid chromatography.

strategies concentrate on the mobile phase because the experiments are easier to automate and column re-equilibration following changes in mobile-phase composition usually occurs within 5–10 column volumes.

3.6.4 Reversed-phase liquid chromatography of ionic compounds

The discovery that ionic solutes can be separated on reversed-phase supports (Horváth et al., 1977a,b; Nahum and Norváth, 1981; Bij et al., 1981; Foley, 1987) was probably one of the most important events in the development of modern liquid chromatography because the majority of compounds of interest to the pharmaceutical, biomedical, biochemical and biological sciences are ionizable. The effects of the nature of the mobile phase and the stationary phase on the retention of ionic compounds are essentially the same as for neutral compounds. Additionally, the retention of ionic compounds is also influenced by secondary equilibria such as dissociation and ion pairing. This significantly increases the flexibility of reversed-phase liquid chromatography. Previously, ion-exchange chromatography was the only viable option for the separation of ionic compounds (see section 3.7).

Two basic approaches have been described to enhance the affinity of ionic compounds for the stationary phase in reversed-phase systems. The first approach, sometimes inappropriately referred to as 'ionic suppression' involves the adjustment of the pH mobile phase so that the compound of interest is in the neutral or unionized form. The second approach to enhance the retention of ionic compounds on reversed-phase supports involves the application of ion-pair extraction techniques.

3.6.4.1 Effects of pH on the retention of ionizable compounds.

Figure 3.9 shows the various equilibria contributing the retention of ionizable compounds in reversed phase systems. In general, the retention of a neutral form of an ionizable compound will be greater than the retention of the ionized form. Therefore, maximum retention will occur at $pH < (pK_a - 2)$ for weak acids, at $pH > (pK_a + 2)$ for weak bases, and at $pH = pI$ for amphoteric substances. It also follows that retention will be sensitive to changes in pH only when $pH = (pK_a \pm 2)$.

The retention of a weak acid in reversed-phase systems can be related to the fractions of the neutral, HA, and conjugate base, A^-, forms present in the mobile phase (Horváth et al., 1977a; Foley, 1987), according to

$$k' = k'_{HA} f_{HA} + k'_{A^-} f_{A^-} \qquad (3.17)$$

where k'_{HA} and k'_{A^-} are the capacity ratios of the neutral and conjugate base form, respectively. The values of k'_{HA} and k'_{A^-} may be related to their distribution constants, K_{HA} and K_{A^-}, (Figure 3.9) by equations (3.18) and (3.19), respectively.

$$k'_{HA} = K_{HA} \frac{V_s}{V_m} \qquad (3.18)$$

$$k'_{A^-} = K_A \frac{V_s}{V_m} \qquad (3.19)$$

Substituting for the dissociation constants into equation (3.17) gives equation (3.20), which relates retention to the hydrogen ion concentration:

$$k' = \frac{k'_{HA}[H^+] + k'_{A^-}K_a}{[H^+] + K_a} \qquad (3.20)$$

The appropriate expressions for the retention of weak bases and amphoteric substances are given in equations (3.21) and (3.22), respectively.

$$k' = \frac{k'_{HB^+}[H^+] + k'_B K_a}{[H^+] + K_a} \qquad (3.21)$$

$$k' = \frac{k'_{H2A^+}[H^+]^2 + k'_{HA^\pm}[H^+]K_{a,1} + k'_{A^-}K_{a,1}K_{a,2}}{[H^+]^2 + [H^+]K_{a,1} + K_{a,1}K_{a,2}} \qquad (3.22)$$

Figure 3.11 shows the theoretical relationships between k' and pH (equations (3.20)–(3.22)) for those ionizable organic compounds that are most commonly encountered in the life sciences: carboxylic acids (pK_a typically 2–6), primary, secondary and tertiary amines (pK_a 7–10) and amino acids (pK_{a1} 2–5, pK_{a2} 8–10). Figure 3.11 shows that selectivity and resolution of ionic compounds having different dissociation constants may be controlled by the pH of the mobile phase. However, it is also clear from Figure 3.10 that those compounds whose retention on silica-based columns will be most effected by the pH of the mobile phase are carboxylic acids and amino acids, because the usable pH range for silica-based columns is 2.5–7.5. In contrast, the retention of most organic amines on silica-based columns will be independent of the pH of the mobile phase because pH values greater than 8 are generally required to generate the neutral form. The inability to generate the neutral form of an amine in reversed-phase systems using silica-based columns has led to the development of polymeric phases such as those based on polystyrene-divinylbenzene which have a much wider range of pH stability (1–13).

3.6.4.2 Reversed-phase ion-pair liquid chromatography. Reversed-phase ion-pair chromatography is an alternative approach for controlling the retention of ionic compounds. This approach is particularly useful for the separation of amines on silica-based columns and it has had a profound effect on the analysis of this class of compounds. In particular, the combination of reversed-phase ion-pair liquid chromatography and electrochemical detection revolutionized the analysis of neurotransmitters in the brain (Tomlinson *et al.*, 1978 and refs. therein).

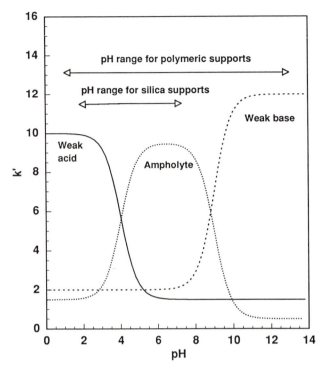

Figure 3.10 Theoretical relationships between k' and mobile phase pH. The curves have been simulated according to following relationships: weak acids, equation (3.20), with $k'_{HA} = 10$, $k'_{A^-} = 1.5$, $pK_a = 4$; weak bases, equation (3.21), with $k'_{HB^+} = 2$, $k'_{B^-} = 12$, $pK_a = 9$; ampholytes, equation (3.22) with $k'_{H_{2A^+}} = 1.5$, $k'_{HA^\pm} = 9.5$, $k'_{A^-} = 0.5$, $pK_1 = 4$, $pK_2 = 2$.

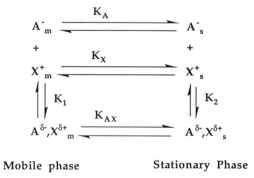

Figure 3.11 Equilibria involved in reversed-phase ion-pair liquid chromatography.

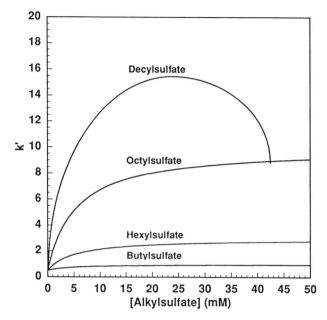

Figure 3.12 Relationships between k' for epinephrine and the concentration of alkyl sulfates in the mobile phase taken from the data of Horváth and co-workers. See Horváth (1977b) for experimental details. The curves for butyl-, hexyl- and octylsulfate have been simulated according to equation (3.25) and the following parameters: (a) butyl sulfonate: $k'_{A^-} = 0.5$, $k'_{AX} = 1$, $K_1 = 200 \, M^{-1}$; (b) hexyl sulfonate: $k'_{A^-} = 0.5$, $k'_{AX} = 3$, $K_1 = 200 \, M^{-1}$; (c) hexyl sulfonate: $k'_{A^-} = 0.5$, $k'_{AX} = 10$, $K_1 = 200 \, M^{-1}$.

In reversed-phase ion-pair liquid chromatography, the pH of the mobile phase is generally adjusted so that the compound or compounds of interest are in their ionized form and retention is manipulated by the addition of an oppositely charged ion-pairing agent to the mobile phase. Quaternary ammonium salts are generally used for the ion-pair chromatography of anions and alkylsulfates or alkylsulfonates are generally used for the separation of cations. The distinction has been made between short alkyl chain ion-pairing agents such as sodium octane sulfonate and long-chained surfactants such as sodium dodecyl(lauryl) sulfate or hexadecyltrimethylammonium bromide (also known as cetyltrimethylammonium bromide or cetrimide). Based on the absence of surfactant properties, symmetrical ion-pairing agents such as tetrabutylammonium ions are usually included with the small, non-surfactant type of ion-pairing agents. Reversed-phase ion-pair liquid chromatography with long-chained ion-pairing agents is sometimes referred to as 'surfactant' (Tomlinson *et al.*, 1978) or 'soap chromatography' (Knox and Laird, 1976). (The latter term is not strictly accurate because soaps are alkaline metal salts of long-chained fatty acids.)

Retention in reversed-phase ion-pair chromatography is dependent upon the concentration and the hydrophobicity (alkyl chain length) of the ion-pairing agents (Figures 3.11 and 3.12). Two mechanisms have been proposed for the retention of solutes in reversed-phase ion-pair liquid chromatography. The first model assumes that the ion-pairs formed in the mobile phase are distributed between the stationary phase and the mobile phase. The second dynamic ion-exchange model assumes that analyte interacts with a monolayer of the ion-pairing agents adsorbed onto the surface of the stationary phase. However, the complete thermodynamic cycle shown in Figure 3.11 demonstrates that both mechanisms are possible even in the presence of a monolayer of adsorbed ion-pairing agent on the surface of the stationary phase. Furthermore, modelling of the relationship between retention (k') and the concentration of ion-pairing agents results in mathematical expression of the same form for both mechanisms.

Using for the moment the ion-pair distribution model (Horváth *et al.*, 1997b), the overall retention of ionized solute may be related to the fractions of the ion-pair free form of the analyte (Horváth *et al.*, 1977b; Foley, 1987):

$$k' = (K_{A^-} f_{A^-} + K_{AX} f_{AX}) \frac{V_s}{V_m} \tag{3.23}$$

$$= \frac{(K_{A^-} + K_1 K_{AX}[X]_m) V_s}{(1 + K_1[X]_m) V_m} \tag{3.24}$$

where K_1 is the ion-pair formation constant, and K_{A^-} and K_{AX} are the distribution coefficients of the anion, A^-, and the ion-pair, AX, respectively. The distribution coefficients may be replaced by the analogous capacity ratios to give

$$k' = \frac{(k'_{A^-} + K_1 k'_{AX}[X]_m)}{(1 + K_1[X]_m)} \tag{3.25}$$

which predicts a hyperbolic relationship between k' and the concentration of ion-pairing agent in the mobile phase. Ion-pair formation constants for small ion-pairs in aqueous solution are generally of the order 10^2 to $10^3 \, M^{-1}$ (Tomlinson *et al.*, 1979) and are relatively independent of solute hydrophobicity. Thus it follows that the increase in retention occurring with increased hydrophobicity (chain length) (Figure 3.12) arises from the increase in the distribution coefficient of the ion-pair (K_{AX}).

If one assumes the dynamic ion-exchange model, then equation (3.26) is obtained, which is identical in form to equation (3.24), demonstrating that retention data alone cannot distinguish these two models of retention and that more detailed studies are necessary (Horváth *et al.*, 1977b).

$$k' = \frac{(K_{A^-} + K_2 K_{X^-}[X]_m) V_s}{(1 + K_2[X]_m) V_m} \qquad (3.26)$$

In this model, retention increases with increasing stationary-phase uptake of the ion-pairing agents (K_X), which has been shown to be related to chain length. Of course, if the ion-pairing agent does not absorb onto the stationary phase the dynamic ion-exchange mechanism cannot occur and the mechanism must involve ion-pairing partitioning. It follows that low concentrations of small, hydrophilic ion-pairing agents in the mobile phase favor the ion-pair model and higher concentrations of more hydrophobic ion-pairing agents favor the dynamic ion-exchange mechanism.

The curves in Figure 3.12 show the hyperbolic relationships between the k' value of epinephrine (adrenaline) and the concentration of butyl- (C_3), hexyl- (C_6) and octyl- (C_8) sulfonates and $[X]_m$ obtained by Horváth and co-workers. This group also demonstrated that more complex, parabolic relationships (Figure 3.12) and sigmoidal relationships (not shown) between k' and $[X]_m$ can occur when using more hydrophobic ion-pairing agents. Sigmoidal relationships between k' and $[X]_m$ can occur as a result of depletion of the ion-pairing agent in the mobile phase, which can arise, in turn, if the concentration of ion-pairing agent in the mobile phase is not substantially in excess of the concentration of the analyte. Therefore, the concentration of ion-pairing agent in the mobile phase should be at least 1 mM. Parabolic relationships between k' and $[X]_m$ (Figure 3.12) probably arise from mixed mechanism of retention in which all the equilibria shown in Figure 3.11 play a significant role in the overall retention. In addition, the presence of micelles in the mobile phase can also contribute to a decrease in retention with increasing concentration of ion-pairing agent (see section 3.6.5).

3.6.4.3 Residual silanol effects in reversed-phase liquid chromatography. Residual silanol groups on the surface of hydrocarbonaceous stationary phases may represent an additional site for the retention of solutes. The silanol group is weakly acidic with a pK_a value ranging between 4 and 6 depending on its microscopic environment. Interactions with the neutral form of the silanol group, which predominates at low pH, may occur through hydrogen bonding with analytes containing a lone pair of electrons. If these types of interactions contribute to retention then the stationary phase can be considered to have both reversed-phase and normal-phase characteristics. It will be recalled (see section 3.6.1) that retention in reversed-phase liquid chromatography decreases with increasing concentration of organic solvent in the mobile phase. In contrast, water is the strongest solvent in normal-phase liquid chromatography and the silanophilic contributions to retention will increase with decreasing amounts of water in the mobile phase. Consequently, when both silanophilic and solvophobic interactions contribute to retention, the value of k' will pass through a minimum value with increasing concentration of organic modifier in the mobile phase. If both the silano-philic and the solvophobic interactions are both very strong then the analyte will be irreversibly retained by the column, irrespective of the type and the concentration of organic modifier in the mobile phase. At higher pH values electrostatic interactions between the anionic form of the silanol group and cationic analytes, most notably aliphatic amines, can contribute to the overall retention. Sometimes advantage can be taken of these secondary interactions to optimize a particular separation. However, interactions between protonated amines and residual silanols may lead to poor reproducibility of retention times and asymmetrical peak shapes.

One approach to solving the problem of residual silanol interactions has involved improvements in the synthetic procedures for the production of hydrocarbonaceous stationary phases. One synthetic approach for the elimination of residual silanol groups involves the reaction of the bonded phase with a small silylating reagent such as trimethylchlorosilane which is presumed to have easier access to silanol groups than bulkier, long-chained chlorosilanes. An alternative, synthetic approach involves surface polymerization of the stationary phase, which is believed to reduce the accessibility of surface silanol groups to polar analytes in the mobile phase. Stationary phases produced by the former method are often referred to as 'end-capped' and stationary phases produced by the latter method are sometimes called 'base-deactivated.'

An alternative to the use of end-capped or base-deactivated columns involves the addition of masking agents to the mobile phase (Riley, 1984; Riley and James, 1986). Weak silanol interactions can sometimes be masked by reducing the pH of the mobile phase. Stronger silanol interac-tions can be masked by the addition of cations to the mobile phase. Ion-pairing agents, such as a quaternary ammonium ions, are commonly used as silanol-masking agents at concentrations of 5–50 mM. However, it should be noted that higher concentrations (100–500 mM) of metal ions such as Li^+, Na^+, K^+ or Ca^{2+} may also be effective in reducing un-wanted silanol interactions (Crouch et al., 1988).

A number of so-called 'double ion-pair' methods have been described for the analysis of hydrophobic amines in which the mobile phase con-tains both a quaternary ammonium ion and an alkyl sulfate or sulfonate. At first glance, this combination of mobile phase additives is counter-intuitive because one would expect the effect of the anionic and cationic additives to cancel. However, the combination of cationic masking agents to reduce peak tailing and an anionic ion-pairing agent to enhance reten-tion is sometimes necessary for the reversed-phase separation of hydro-phobic amines.

3.6.4.4 Retention and selectivity control for ionic molecules in reversed phase liquid chromatography. Figures 3.10 and 3.12 show that the required retention for a given ionic analyte can be achieved by varying the pH of the mobile phase, as well as the chain length and the concentration of ion-pairing agent in the mobile phase. However, ion-pairing agents are expensive and ion-pair chromatography should only be considered if optimization strategies investigating the effects of pH, organic modifier type and concentration have proved unsuccessful. Thus if the compounds to be separated differ in pK_a values then the solvent strength should first be adjusted to give a k' value between 2 and 10 for the analytes of interest. The pH of the mobile phase should then be investigated to achieve the best separation. Foley (1987) has shown that the optimum pH, pH_{opt} for two closely eluting weak acids is given by

$$pH_{opt} = pK_{a,av} + 0.5 \log \left(\frac{k'_{HA,av}}{k'_{A^-,av}} \right) \qquad (3.27)$$

where $pK_{a,av}$ is the average of the two pH values and $k'_{HA,av}$ and $k'_{A^-,av}$ are the average k' values for the HA and A^- forms of the two compounds, respectively.

If ion-pair chromatography is indicated, then concentrations of ion-pairing reagent in the range 1–50 mM should be employed. Retention will be either independent of concentration or decrease with increasing concentration of ion-pairing agent above 25–50 mM. If cost is a consideration, then low concentrations of longer chain length ion-pairing agents are preferred to higher concentrations of more hydrophilic reagents. However, the solubility of the ion-pairing reagent in the mobile phase may preclude the use of higher concentrations.

The relationship between selectivity, the nature of the ion-pairing reagent and its concentration merit some consideration. Clearly, ion-pair chromatography is a useful technique for increasing the retention of ionized solutes of different charges in a selective manner. However, the effects of ion-pairing on selectivity of analytes of the same charge are less obvious. It is reasonable to assume that the optimum concentration of ion-pairing agent, [X] may be given by (Foley, 1987):

$$p[X]_{m,opt} = pK_{1,av} + 0.5 \log \left(\frac{k'_{AX,av}}{k'_{A^-,av}} \right) \qquad (3.28)$$

where pK is the average of the negative logarithm of ion-pair association constants and $k'_{AX,av}$ and $k'_{A^-,av}$ are the average k' values for the AX and A^- forms of the two compounds, respectively. This assumes that the selectivity arises from differences in ion-pair formation constant $(K_{1,2}/K_{1,1})$, and the ion-pair distribution constants $(K_{ax,2}/K_{AX,1})$. The selectivity factor for two adjacent peaks, 1 and 2, that are both retained by reversed-phase ion-pair liquid chromatography, α (see equation (2.41)) is given by

$$\alpha = \frac{k'_2}{k'_1} \tag{2.41}$$

Substitution of equation (3.25) for k'_1 and k' into equation (2.41) gives

$$\alpha = \frac{(K_{A,2^-} + K_{1,2}K_{AX,2}[X]_m)(1 + K_{1,1}[X]_m)}{(K_{A,1^-} + K_{1,1}K_{AX,1}[X]_m)(1 + K_{1,2}[X]_m)} \tag{3.29}$$

which predicts that selectivity depends upon differences in the ion-pair formation constants, $K_{1,1}$ and $K_{1,2}$, as well as differences in the distribution coefficients of the two ion-pairs, $K_{AX,1}$ and $K_{AX,2}$. With the exception of compounds where the ionized groups are hindered, ion-pair formation constants of small ion-pairs are relatively insensitive to structural changes. If it is also assumed that $[X] >> k'_{A^-}$ in the numerator and the denominator of equation (3.29), then equation (3.29) simplifies to

$$\alpha = \frac{k'_{AX,2}}{k'_{AX,1}} \tag{3.30}$$

which states that selectivity for small ion-pairs depends only on the distribution coefficients of the two ion-pairs, $K_{AX,1}$ and $K_{AX,2}$. Unfortunately, the ion-pair distribution coefficient is determined by solute hydrophobicity, which is the same factor that governs selectivity in the absence of the ion-pairing agents, i.e.

$$\alpha = \frac{(K_{A,2})}{(K_{A,1})} = \frac{(K_{AX,2})}{(K_{AX,1})} \tag{3.31}$$

Tomlinson *et al.* (1979a) have shown that the ion-pair formation constants of large ion pairs of more than 25 carbon atoms are related to the total number of carbon atoms in the ion pair. Thus for larger ion pairs the situation is more complex because both differences in the ion-pair formation constant and the distribution coefficient of the ion pair will contribute to selectivity; however, both these factors depend on solute hydrophobicity. These theoretical predictions are confirmed by experimental observations and many studies have shown that selectivity in reversed phase ion-pair systems is independent of the chain length of the ion-pairing reagent and its concentration in the mobile phase (Horváth *et al.*, 1977b, Tomlinson *et al.*, 1979). If analytes of the same charge are poorly separated by reversed-phase ion-pair chromatography, then differences in pK_a values should be exploited to improve resolution. In this case, the effects of ion-pairing reagent chain length and concentration and the pH of the mobile phase should be investigated to achieve optimum resolution.

In contrast to the separation of analytes of the same charge, the selectivity of molecules of different charge is very sensitive to the type and the concentration of the ion-pair agent. In addition, the effects of organic modifier on the retention of ion-pairs are essentially the same as for neutral molecules (section 3.6.1). Thus, optimization of ion-pair separation should involve studying the effects of pH, organic modifier type and organic modifier concentration, as well as the effects of ion-pair reagent and concentration. Finally, enhancing the affinity of ionic solutes for the stationary phase by ion-pairing with oppositely charged ions permits the use of stronger solvents systems in the mobile phase. Using stronger solvent systems (i.e. those with higher concentrations of organic modifiers) can be useful in the analysis of biological fluids because it will reduce the retention of very hydrophobic, neutral compounds that would otherwise be strongly retained by the column.

3.6.5   Micellar chromatography

Micellar chromatography is the term specifically reserved for that form of reversed-phase liquid chromatography that employs a micellar mobile phase to separate both neutral and ionic analytes which would otherwise have a very high affinity for the stationary phase (DeLuccia et al., 1985). In this mode of chromatography the analytes partition into the micelles in the mobile phase as well as into the stationary phase. One of the main applications of micellar chromatography is for the direct injection of plasma and other biological fluids. The micelles in the mobile phase solubilize the plasma proteins which thus elute at the solvent front rather than being highly retained or precipitated (DeLuccia et al., 1985).

3.7   Ion-exchange chromatography

When choosing a chromatographic format for the analysis of an ionic compound, ion exchange is generally considered after attempts at developing a reversed-phase or reversed-phase ion-pair method have proved unsuccessful. However, ion-exchange chromatography is the method of choice for the analysis of inorganic ions and it is often preferable to reversed-phase methods for the analysis of small organic ions. Furthermore, most ion-exchange methods employ simple aqueous buffers as mobile phases and are easy to automate. Consequently, the separation step in all commercial ion chromatographs is carried out by ion-exchange chromatography. It is also the method of choice for the separation of amino acids prior to post-column derivatization with ninhydrin, fluorescamine or o-phthlaldehyde/thiol. Ion-exchange chromatography is used extensively for the preparative separation of peptides and proteins because aqueous (non-denaturing) mobile phases are used. Carbohydrates may also be separated in their anionic forms at high pH or as anionic boronate complexes on anion exchangers. The main disadvantage of ion-exchange chromatography is the lower column efficiencies compared with reversed-phase liquid chromatography. However, ion-exchange separations are often performed with gradient elution and elevated temperature, which tend to reduce peak dispersion.

3.7.1 Stationary phase effects in ion-exchange chromatography

Retention in ion-exchange chromatography arises from electrostatic interactions between ions in the mobile phase and oppositely charged ionic functional groups immobilized on the surface of a solid support. Ion-exchangers are divided into anion-exchangers and cation exchangers according to the nature of the functional groups on the surface of the stationary phase. Anion exchangers and cation exchangers are further classified as being either weak or strong according to the dissociation constant of the immobilized functional group. Typically strong anion exchangers (SAX) are immobilized quaternary ammonium ions and strong cation exchangers (SCX) are immobilized sulfonic acids, which are ionized over the complete pH range. In contrast, weak anion exchangers (WAX) are immobilized amines and weak cation exchangers (WCX) are immobilized carboxylic acids that are only ionized at certain pH values.

The first ion-exchanger resins introduced into column–liquid chromatography were pellicular materials in which the stationary phase was attached to the surface of a glass bead. Pellicular ion-exchangers have largely been replaced by ion-exchangers that are covalently attached to either silica gel or some type of porous polymer. The advantage of silica-based supports is the high mechanical strength and their resistance to high pressures compared with the polymer-based materials. The main advantage of polymer-based stationary phases, on the other hand, is their resistance to chemical degradation at high pH values.

3.7.2 Retention mechanisms and mobile phase effects in ion-exchange chromatography

Retention in ion-exchange chromatography is determined by the pH of the mobile phase, ionic strength, temperature and the nature and the concentration of buffer ions in the mobile phase. Organic modifiers such as methanol, acetonitrile or tetrahydrofuran may also be added to the mobile phase because retention of hydrophobic ions may also arise as a result of interactions with the stationary-phase support or the hydrocarbonaceous functional group linking ion-exchange group to the support.

The competition between ions in the mobile phase and adsorbed counter ions on the stationary phase is given by equations (3.32) and (3.33) for anion exchangers and cation exchangers, respectively:

SAX (chloride-ion form)

$$(R - NR_3^+)_s Cl^- + Y^-_m \rightleftharpoons (R - NR_3^+)_s Y^- + Cl^-_m \qquad (3.32)$$

SCX (sodium-ion form)

$$(R - SO_3^-)_s Na^+ + X^+_m \rightleftharpoons (R - SO_3^-)_s X^+ + Na^+_m \qquad (3.33)$$

In follows from equations (3.32) and (3.33) that the ion-exchange constant (K_{EX}) for monovalent ions, A and B is given by (Poole and Poole, 1991):

$$K_{EX} = \frac{[A]_s [B]_m}{[A]_m [B]_s} \qquad (3.34)$$

The ion-exchange constant is sometimes referred to as the ion-exchange selectivity constant because it defines the relative affinity of an ion-exchanger for two ions. Thus the ion-exchange constant forms the basis of an elutropic series of counter ions, which may be used in the development of an ion-exchange method. Affinity of an ion for an ion-exchanger depends on charge and ionic radius. Retention increases with increasing net charge of the solute, but for ions of equal charge, solutions of lithium salts are the weakest solvents on strong cation exchangers and solutions of fluoride salts are the weakest solvents on strong anion exchangers (Snyder and Kirkland, 1979; Poole and Poole, 1991) (Table 3.6). The order of affinity is the same for strong ion exchangers as it is for weak ion exchangers with one important exception. The hydrogen ion has the strongest affinity of all the cations for weak cation exchangers

Table 3.6 Affinities of ions for strong-anion exchangers and strong-cation exchangers

Relative affinity[a]	Cations	Anions
Strongest	Ba^{2+}	$(Citrate)^{3-}$
	Pb^{2+}	SO_4^{2-}
	Ca^{2+}	$(Oxalate)^{2-}$
	Ni^{2+}	I^{-b}
	Cu^{2+}	NO_3^-
	Co^{2+}	Br^- *
	Zn^{2+}	SCN^-
	Mg^{2+}	Cl^{-b}
	Ag^+	$(Formate)^-$
	Cs^+	$(Acetate)^-$
	K^+	OH^-
	NH_4^+	F^{-b}
	Na^+	
	H^+	
Weakest	Li^+	

[a]Taken from Schoenmakers (1982).
[b]Halide salts should not be used in stainless steel systems because they are corrosive.

and the hydroxide ion has the strongest affinity of all anions for anion exchangers.

The ion-exchange constant also determines the elution order for a series of ionic solutes and it may be related to the capacity ratio of an analyte ion, A, in the presence of a counter ion, B. Equation (3.34) may be rearranged to give equation (3.35), which relates the capacity ratio of A to the ion exchange constant for that pair of ions and the concentrations of the counter ion, B, in the mobile phase and the stationary phase

$$k'_A = \frac{1}{K_{EX}} \frac{[B]_s}{[B]_m} \frac{V_s}{V_m} \qquad (3.35)$$

Assuming $[B]_s >> [A]_s$, the concentration of counter ions in the stationary phase is equal to the concentration of ion-exchange sites in the stationary phase, Q. Q is defined as the ion-exchange capacity of the stationary phase:

$$k'_A = \frac{1}{K_{EX}} \frac{Q}{[B]_m} \frac{V_s}{V_m} \qquad (3.36)$$

$$\log k'_A = \log\left(\frac{Q}{K_{EX}} \frac{V_s}{V_m}\right) - \log[B]_m \qquad (3.37)$$

Equations (3.36) and (3.37) predict that retention in ion-exchange chromatography decreases with increasing concentration of counter ions in the mobile phase. Thus retention is optimized by appropriate choice of the type and concentration of electrolytes added to the mobile phase.

The ion-exchange constant does vary from one pair of ions to another; therefore selectivity can be controlled by appropriate choice of the counter ion. However, selectivity is more conveniently manipulated by controlling the pH of the mobile phase and taking advantage of differences in the pK_a values of the analytes to be separated. In contrast with reversed-phase liquid chromatography (see section 3.6.2.1) the retention of weak acids and weak bases will reach a maximum when the compounds are in their ionized forms. Zwitterionic compounds such as amino acids, peptides and proteins can be separated on anion exchangers or cation exchangers.

The ability of ion-exchangers to distinguish ions of different charge is high and the separation of ions of different charges is generally conducted under gradient-elution conditions. Solvent gradients can be generated in ion-exchange chromatography by continuously increasing the concentration of a given counter ion, or in stepwise fashion by abruptly changing the pH, the nature of the counter ion or its concentration in the mobile phase. Temperature gradients have also been used to decrease the affinity of ions for the stationary phase.

3.8   Size-exclusion chromatography

Separation in reversed-phase liquid chromatography, normal-phase liquid chromatography and ion-exchange chromatography arises from different interactions of the solutes with the mobile phase and the stationary phase. In contrast, separations in size-exclusion chromatography arise from differences in molecular size and the ability of different molecules to penetrate the pores of the stationary phase to different extents. Historically size-exclusion chromatography has been divided into gel-filtration chromatography which uses hydrophilic dextrans for the separation of water-soluble polymers and gel-permeation chromatography, which is used for the separation of synthetic, organic polymers. This distinction is arbitrary because both techniques employ the same separation process.

Size-exclusion chromatography is used extensively for the preparative separations of macromolecules of biological origin as well as for the purification of synthetic-organic polymers. It is particularly well suited to the isolation of fragile peptides and proteins because the gentle elution conditions tend not to denature the analytes. The main analytical applications of size-exclusion chromatography are for determination of the molecular weight of single macromolecules and for the determination of molecular weight distributions of polydispersed polymers. Because of the poor peak capacity of size-exclusion chromatography, this technique is less useful for the quantitative determinations of specific macromolecules such as a specific protein in complex samples. In that case reversed-phase liquid chromatography or ion-exchange chromatography are preferred. However, size-exclusion chromatography can be used as a primary clean-up step in multi-dimensional separation of low molecular weight analytes from complex biological samples.

3.8.1   Separation mechanism in size-exclusion chromatography

Separation in size-exclusion chromatography requires careful matching of the pore size of the stationary phase material with the size of the molecules to be separated. Small molecules in a sample will be able to penetrate all the pores of the stationary phase and will elute with an elution volume, V_e, which is equal to the void volume of the column, V_m. Very large molecules will be excluded from all the pores of the stationary phase and will elute with an elution volume, which is equal to the interstitial volume of the liquid between the particles, V_i. Molecules of intermediate size will be able to penetrate some but not all of the pores and will elute with an elution volume that is between the interstitial volume and the void volume of the column. The void volume is the total volume of the liquid in the column and is related to the interstitial volume and the pore volume, V_p by equation (3.38):

$$V_m = V_o = V_i + V_p \tag{3.38}$$

The total volume of the column, V_{tot}, and its porosity, ε, are given by equation (3.39) and (3.40), respectively:

$$V_{tot} = V_s + V_m \tag{3.39a}$$

$$= V_s + V_i + V_p \tag{3.39b}$$

$$= \pi d_c^2 L/4 \tag{3.39c}$$

$$\varepsilon = \frac{V_m}{V_{tot}} = \frac{V_m}{\pi r_c^r L} \tag{3.40}$$

It follows that the elution volume of a solute in size-exclusion chromatography is given equation (3.41):

$$V_e = V_i + K_{SEC} V_p \tag{3.41}$$

where the size-exclusion constant, K_{SEC}, has values between 0 and 1.

The main disadvantage of size-exclusion chromatography is its low peak capacity, which arises from the small elution volumes of the peaks. In addition because peak dispersion is small, extra-column contributions to band broadening in size-exclusion chromatography have to be kept to a minimum. The relative contribution of extra-column band broadening can be minimized by the use of small particles ($d_p = 3$–$10\,\mu m$) and long ($L = 50$–$100\,cm$), wide-bore ($d_c = 8$–$10\,mm$) columns.

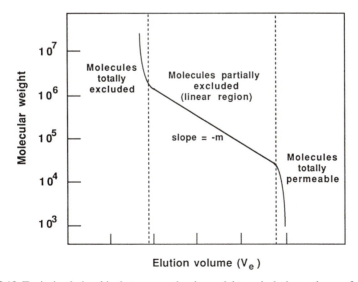

Figure 3.13 Typical relationship between molecular weight and elution volume of polymers on a size-exclusion column. Adapted from Miller (1988).

3.8.2 Stationary phase and mobile phase effects in size exclusion chromatography

Unlike the other modes of chromatography, interactions with the stationary phase surface are to be avoided in size-exclusion chromatography. Thus the important criteria for the choice of a stationary phase in size-exclusion chromatography are the pore-size distribution, and compatibility with the mobile phase and the analytes of interest. Any interaction with the stationary phase will lead to retention and an increase in elution volume. The reduction of interactions of the analyte with the surface is particularly important when using size-exclusion chromatography for molecular weight determinations because an increase in elution volume will result in a reduction in the apparent molecular weight (Figure 3.13). The criteria for the choice of an appropriate mobile phase for size-exclusion chromatography include solubility of the analytes, viscosity, chemical compatibility with the sample, compatibility with the stationary phase and compatibility with the detector.

Stationary phases for size-exclusion chromatography can be conveniently divided into rigid polymers and soft gels. Soft gels such as crosslinked dextran (e.g. Sephadex®) that swell when placed in water are very compressible and therefore unsuitable for use in pressurized systems. Rigid size-exclusion phases that can be used in pressurized columns are made of either silica gel, derivatized silica gel or polymeric materials such as polystyrene. Rigid size-exclusion supports have been described for the separation of compounds with RMM less than 1000. However, the other modes of chromatography that rely on specific stationary-phase and mobile-phase interactions are much more selective than size-exclusion chromatography for the separation of small molecules.

For molecules that are partially excluded from the pores of the stationary phase, the elution volume is related to the relative molecular mass, RMM, of the polymer by equation (3.42) (Poole and Poole, 1991; Miller, 1988a):

$$\log \mathrm{RMM} = I - m V_e \qquad (3.42)$$

where I and m are the intercept and slope coefficients, respectively (Figure 3.13). For the determination of the relative molecular mass of a single polymer, calibration curves are established using polymers of known relative molecular mass. Proteins such as cytochrome C (RMM = 12 500), bovine serum albumin (RMM = 68 000), and γ-globulin (RMM = 167 000) are used for the relative molecular mass calibration of water-soluble biopolymers, and synthetic polymers such as polystyrenes are used for the analysis of organic polymers. Polydispersed polymers will elute from a size-exclusion column as a single broad peak and computer analysis of the peak shape may be used to determine molecular weight distribution functions such as the weight-average molecular weight, the number-average molecular weight and the z-average molecular weight. The choice of stationary phase and calibration standards with the size distribution appropriate for the analyte of interest is important because the range of RMMs that can be separated with a single size-exclusion column is only about 1.5 log units (Figure 3.13). Several size-exclusion columns may be linked series to increase the range of RMMs that can be determined. However, high pressure drops will limit the total number of columns that may be coupled together.

Bibliography

Bij, K.E., Horváth, C., Melander, W.R. and Nahum, A. (1981) Surface silanols in silica-bonded hydrocarbonaceous stationary phases. II. Irregular retention behavior and effect of silanol masking. *J. Chromatogr.* **203**, 65–84.

Crouch, F.W., Riley, C.M. and Stobaugh, J.F. (1988) The high performance liquid chromatography of bis-[1,2-bis-(diphenylphosphino)ethane]gold (I) chloride [Au(dppe₂)]Cl, a novel antineoplastic agent. *J. Chromatogr.* **448**, 333–343.

DeLuccia, F.J., Arunyanart, M. and Cline Love, L.J. (1985) Direct serum injection with micellar liquid chromatography for therapeutic drug monitoring. *Anal. Chem.* **57**, 1564–1568.

Foley, J.P. (1987) The role of secondary equilibria in modern liquid chromatography. *Chromatography* **June**, 43–52.

Hennion, M.C., Picaud, C. and Caude, M. (1978) Influence of the number and the length of alkyl chains on the chromatographic properties of hydrocarbonaceous bonded phases. *J. Chromatogr.* **166**, 21–35.

Horváth, C. and Melander, W. (1977) Liquid chromatography with hydrocarbonaceous bonded phases; theory and practise of reversed phase chromatography. *J. Chromatogr. Sci.* **15**, 393–404.

Horváth, C., Melander, W. and Molnár, I. (1976) Solvophobic interactions in liquid chromatography with nonpolar stationary phases. *J. Chromatogr.* **125**, 129–156.

Horváth, C., Melander, W. and Molnár, I. (1977a) Liquid chromatography of ionogenic substances with nonpolar stationary phases. *Anal. Chem.* **49**, 142–153.

Horváth, C., Melander, W., Molnár, I. and Molnár, P. (1977b) Enhancement of retention by ion-pair formation in liquid chromatography with nonpolar stationary phases. *Anal. Chem.* **49**, 2295–2305.

Karger, B.L. and Geise, R.W. (1978) Reversed phase liquid chromatography and its application to biochemistry. *Anal. Chem.* **50**,1048A–1073A.

Karger, B.L., Gant, J.R., Hartkopf, A. and Weiner, P.H. (1976) Hydrophobic effects in reversed-phase liquid chromatography. *J. Chromatogr.* **128**, 65–78.

Knox, J.H. and Laird, G.R. (1976) Soap chromatography – a new high-performance liquid chromatographic technique for the separation of ionizable materials. Dyestuff intermediates. *J. Chromatogr.* **122**, 17–34.

MacBlane, D., Kitagawa, N. and Benson, J.R. (1987) A C18-derivatized polymer for reversed-phase chromatography. *Am. Lab.* **Feb.**, 134–138.

Majors, R.E. (1980) Practical operation of bonded-phase columns in high-performance liquid chromatography, in *High Performance Liquid Chromatography. Advances and Perspectives*, Vol. 1, ed. C. Horváth, Academic, New York, p. 82.

McCormick, R.M. and Karger, B.L. (1980) Role of organic modifier sorption on retention phenomena in reversed-phase liquid chromatography. *J. Chromatogr.* **199**, 259–273.

Miller, J.M. (1988) *Chromatography. Concepts and Contrasts*, Wiley–Interscience, New York, p. 188.

Nahum, A. and Horváth, C. (1981) Surface silanols in silica-bonded hydrocarbonaceous stationary phases. I. Dual retention mechanism in reversed-phase chromatography. *J. Chromatogr.* **203**, 53–63.

Poole, C.F. and Poole, S. K. (1991) *Chromatography Today*, Elsevier, New York.

Riley, C.M. (1984) The analysis of metoclopramide in plasma by reversed phase ion-pair high performance liquid chromatography. *J. Pharm. Biomed. Anal.* **2**, 81–89.

Riley, C.M. and James, M.O. (1986) Determination of ketoconazole in the plasma, liver, lungs and adrenals of the rat by high performance liquid chromatography. *J. Chromatogr.* **377**, 287–294.

Schoenmakers, P.J., Billiet, H.A.H. and de Galan, L. (1982) The solubility parameter as a tool to understanding liquid chromatography. *Chromatographia* **15**, 205–214.

Schomburg, G. (1988) Stationary phases in high performance liquid chromatography. *LC-GC* **6**, 36–50.

Scott, R.P.W. and Kucera, P. (1979) Solute–solvent interactions on the surface of silica gel. II. *J. Chromatogr.* **171**, 3748.

Sinanoglu, O. (1967) Intermolecular forces in liquids. *Advances in Chemical Physics*, Vol. 12, Wiley, New York, pp. 283–325.

Snyder, L.R. (1968) *Principles of Adsorption Chromatography*, Marcel Dekker, New York.

Snyder, L.R. (1983) Mobile phase effects in liquid–solid chromatography. Importance of adsorption site geometry, adsorbate delocalization and hydrogen bonding. *J. Chromatogr.* **255**, 3–26.

Tanaka, N., Goodell, H. and Karger, B.L. (1978) The role of organic modifiers on polar group selectivity in reversed-phase liquid chromatography. *J. Chromatogr.* **158**, 233–248.

Tomlinson, E., Davis, S.S. and Mukhayer, G.I. (1979) in *Solution Chemistry of Surfactants*, Vol. 1, ed. K.L. Mittal, Plenum, London, pp. 3–43.

Tomlinson, E., Riley, C.M. and Jefferies, T.M. (1979) Functional group behaviour in RP-HPLSC, using surface active pairing ions. *J. Chromatogr.* **185**, 197–224.

Tomlinson, E., Riley, C.M. and Jefferies, T.M. (1978) Ion-pair high performance liquid chromatography. *J. Chromatogr.* **159**, 315–358.

Wells, J.I. (1988) *Pharmaceutical Preformulation. The Physicochemical Properties of Drugs*, Ellis Horwood, Chichester.

4 Support materials and solvents

P. HAMBLETON

4.1 Adsorption chromatography

In HPLC, the stationary phase is normally packed into a stainless steel tube through which the liquid mobile phase is pumped. The practical aspects of such a system, as well as the theoretical principles of liquid chromatography (Chapter 2) place considerable demands on any material which is to be used as a stationary phase. To minimise band-broadening and thereby to improve column efficiency, it is necessary to use small diameter particles for the column packing (Figure 4.1). This presents a problem as there are technical difficulties in producing very small diameter particles of a reasonably uniform size. As HPLC has progressed the particle size of stationary phases has steadily reduced. In the late 1970s HPLC columns were most often packed with $10 \mu m$ (or even $20 \mu m$) particles. Currently, most HPLC separations are carried out with $5 \mu m$ diameter packing materials with materials of $3 \mu m$ diameter also being readily available. Particle sizes as small as $1 \mu m$ may be used, although their application is somewhat more specialised and the normal range of particle sizes used in analytical applications with conventional columns is between $3 \mu m$ and $10 \mu m$.

The desirability of small diameter particles from a column efficiency viewpoint has ramifications for the operating pressure used to pump the mobile phase through the column. As the particle size decreases, so the pressure required to maintain any given flow rate will increase. With conventional columns, and particle sizes as mentioned above, operating pressures vary from approximately 250 psi to 3000 psi depending on factors such as the viscosity of the mobile phase and the flow rate. This places a further demand on the column packing since it must be able to withstand these operating pressures without any structural damage.

In addition to the requirements for the production of small, uniform, porous and rigid particles, a potential HPLC stationary phase should be readily available in a pure form and be chemically resistant to the solvents used as mobile phases (e.g. ideally, it should be stable at all pH values). As is often found in HPLC, there is no single material which fulfils all the criteria, but a variety of supports are used, each with its own virtues and failings.

The material which is most commonly used for HPLC column packing is silica. It may be used either unmodified or after chemical derivatisation

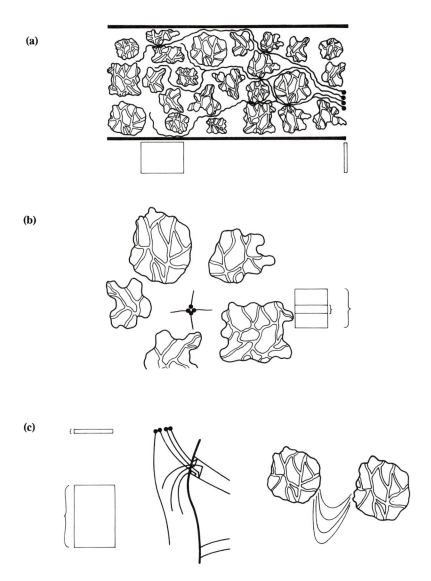

Figure 4.1 Band-broadening processes in porous irregular microparticles. (a) eddy diffusion – analyte molecules take different routes to circumnavigate the particles. They also move more quickly through wide channels than through narrow channels. (b) diffusion in the mobile phase. The short bracket indicates initial band width, the long bracket indicates final band width. (c) mass transfer. On the left is shown mass transfer in stagnant mobile phase in pores, and that due to the adsorption/desorption process. The narrow band represents initial band width, the broad band final band width. On the right is shown mobile phase mass transfer caused by laminar flow.

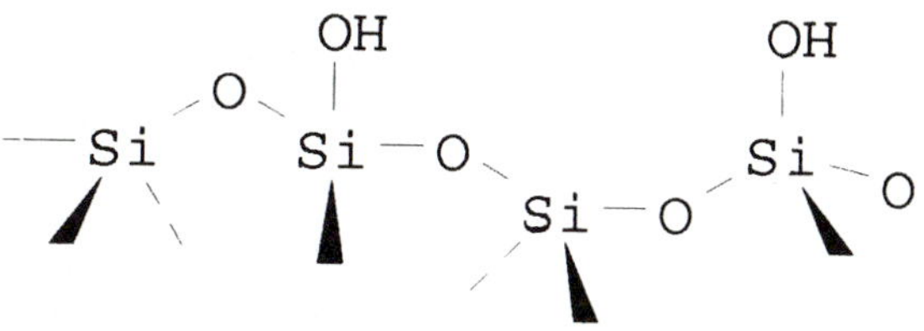

Figure 4.2 Silica surface.

of the silanol groups, allowing the surface chemistry to be altered to accommodate different modes of chromatography.

It is most often prepared by acid hydrolysis of sodium silicate followed by emulsification in an alcohol water mixture and subsequent condensation to give solid silica gel. This is then washed and dried for use as HPLC column packing. The exact conditions under which these procedures are carried out (e.g. pH, catalysts, temperature) will affect the properties of the resulting material. The most important qualities with regard to the chromatographic performance of the gel are the average particle size, the particle shape, the specific surface area and the pore size. Other factors which are also important are the pH of the gel surface, the number of active silanol groups and the presence of metal ions.

Chemically, silica gel is a polymer composed of tetrahedral silicon atoms connected through oxygen atoms (siloxane, Si–O–Si) with silanol (Si–OH) groups present at the surface (Figure 4.2).

Silica gel may be made in either spherical or non-spherical forms depending on the conditions used for condensation and solidification. It is generally considered that spherical forms give superior column packing properties to the non-spherical forms, although this is difficult to demonstrate in practice. As the particle size of the column packing decreases, the superiority of spherical materials becomes more apparent and therefore spherical forms are often used for silica particles of 3 μm diameter or less. The manufacture of spherical silica is much more difficult than that of irregular shaped silica which therefore leads to a higher price, no doubt contributing greatly to the loyalty of chromatographers to non-spherical silica.

Commercially prepared silica is also characterised by the size of the pores. For the separation of low molecular weight compounds it is recommended that the pore width should be at least 5 nm in diameter. For the separation of macromolecules, it is more common to use a material with a pore width of at least 30 nm. If the silica has a pore width smaller than these values then the analytes will be wholly or partially excluded from the pores and a form of size exclusion chromatography will result.

The specific surface area of silica is very high, many commercial products having a value of approximate $400 \, \mathrm{m^2 \, g^{-1}}$. This implies that silica generally has a fairly high sample capacity, making it suitable for a wide range of chromatographic applications. All other things being equal, a high specific surface area will increase the retention of the solutes by increasing the capacity factor k'. The different values for the specific surface area arise due to variations in the manufacturing process and will be affected by the number of available silanol groups on the silica surface. The two principal factors which affect this are the surface area of the pores (small pore diameters leading to high specific surface areas) and the proportion of free silanols. During the drying process the silica is activated by heating and thereby removing adsorbed water. Any water on the gel surface is associated with the silanol groups by hydrogen bonding, thus the degree of activation of the silica can be controlled either during the drying process or by subsequent re-hydration. Drying at very high temperatures (400°C to 600°C) leads to an irreversible loss of water from two adjacent silanols to give a siloxane bond (Si–O–Si) and a consequent reduction in surface activity.

The surface of the silica gel may act as an acid, a base or be essentially neutral, depending on the manufacturing process. This proton donor/ acceptor effect may strongly influence the chromatographic behaviour of acids and bases. It can be shown that use of an 'acidic' silica for acids gives a good peak shape whereas using the same material for the separation of basic compounds often gives marked peak tailing with unacceptably high capacity factors. The situation for a 'basic' silica is reversed, i.e. good chromatography is achieved for bases but not for acids.

As the preceding discussion shows, there is much scope for variation in the production of silica for HPLC. This allows for some degree of customisation of the material for a specific use but also leads to difficulties in standardisation. The result of this is that silica gels from different manufacturers may have very different chromatographic properties. Even different batches of silica from the same manufacturer using a standardised process may differ significantly, requiring a change in an established HPLC method in order to restore the desired separation. Some of these differences may be due to changes in the water content, which has a profound affect in adsorption chromatography or in other cases, to the level of trace metals present.

An additional problem with the use of silica is its limited pH stability. Above pH 8, silica becomes increasingly soluble and it is not advisable under normal circumstances to use any silica based packing at pH values less than 2 or greater than 8.

The main use of unmodified silica in HPLC is as an adsorbent in normal phase chromatography where the mobile phase is less polar than the stationary phase, and in this mode a non-aqueous mobile phase is commonly selected. Due to the high polarity of water and its strong affinity for free silanols on the silica surface, the presence of even very small amounts of water in the mobile phase may have a large effect on the retention of solutes. This may lead to considerable variation in capacity factors from day to day due to absorption of water from the atmosphere. The problem can be partly overcome by using anhydrous solvents and then adding a small but constant amount of water before use.

Despite the limitations of unmodified silica as a stationary phase for normal phase chromatography, it remains the most common adsorbent in this field, although there has been a recent renewal of interest in the use of alumina.

Alumina is a basic compound but it can be processed so that it may exhibit neutral or even acidic properties. The surface of alumina interacts with solutes in a different manner to that of silica. The presence of OH^- and O^{2-} at the surface permits strong adsorption of acidic compounds whilst the presence of aluminium ions (Al^{3+}) allows interaction with compounds containing polarisable groups. The basic nature of alumina gives it an advantage over silica for the chromatography of basic compounds which typically adsorb strongly on silica. However, the situation for acidic compounds on alumina is generally rather poor and often irreversible adsorption occurs. The principle reason for selecting alumina over silica is that it offers a different selectivity due to the different surface chemistry, and as it is stable over a wider pH range (from 2 to 12), separations which are enhanced at high pH are more easily carried out on alumina.

Commercial alumina packing materials for HPLC have lower column efficiency values and lower specific surface areas (typically 100–$250 \, m^2 \, g^{-1}$) which leads to a lower sample capacity. Another disadvantage of alumina is that the highly active surface has a tendency to catalyse deterioration reactions, which is particularly important for molecules which undergo base-catalysed degradation, because of the basic nature of the alumina surface.

Other materials have also been used for adsorption chromatography, notably some metal oxides such as zirconium oxide and titanium oxide. Their behaviour has been compared to both silica and alumina, but despite differences in selectivity none have achieved popularity and their importance as HPLC phases remains very minor.

4.2 Chemically bonded silica

One of the main reasons for the widespread use of silica as an HPLC packing material is that it may be used either unmodified or after chemical derivatisation of the silanol groups, allowing the surface chemistry to be altered to suit different modes of chromatography.

Derivatisation may be carried out in three main ways (Figure 4.3). Firstly, by reaction of the silanol with an alcohol, producing an alkoxy

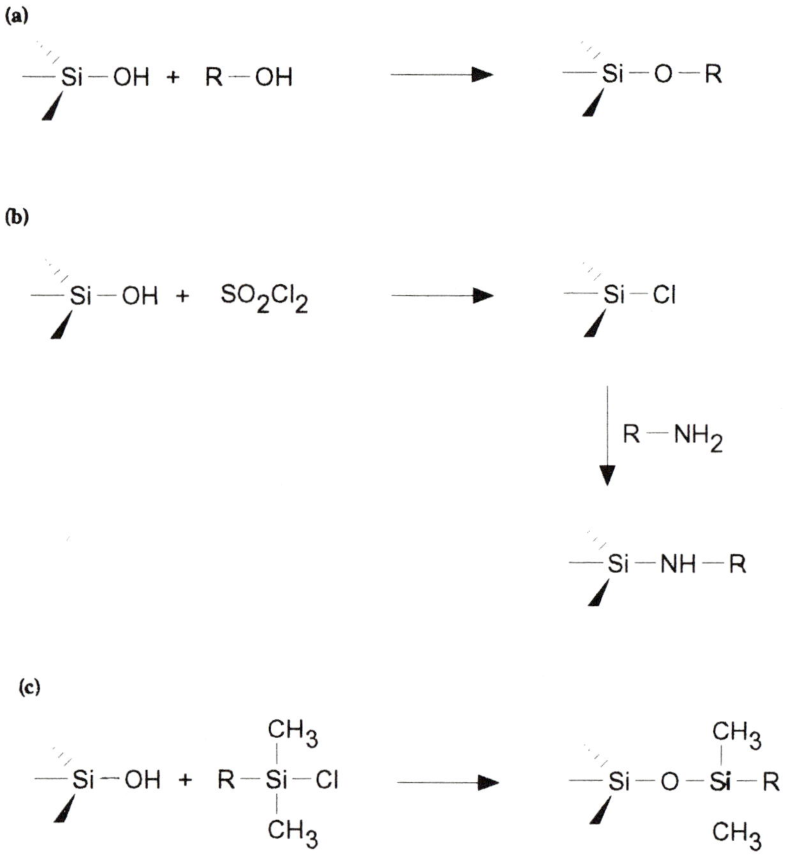

(a)

$$—Si—OH \ + \ R—OH \quad \longrightarrow \quad —Si—O—R$$

(b)

$$—Si—OH \ + \ SO_2Cl_2 \quad \longrightarrow \quad —Si—Cl$$

$$\downarrow R—NH_2$$

$$—Si—NH—R$$

(c)

$$—Si—OH \ + \ R—\underset{\underset{CH_3}{|}}{\overset{\overset{CH_3}{|}}{Si}}—Cl \quad \longrightarrow \quad —Si—O—\underset{\underset{CH_3}{|}}{\overset{\overset{CH_3}{|}}{Si}}—R$$

Figure 4.3 Formation of chemically bonded silica stationary phases.

silane (Figure 4.3(a)). These materials are relatively simple to prepare but are hydrolysed easily and therefore cannot be used with aqueous or alcoholic mobile phases. Secondly, by the production of a chloride using thionyl chloride followed by reaction with an amine to give an alkylaminosilane. These products are more stable to hydrolysis than the alkoxy silanes; however, their preparation is more complex (two stage synthesis) and considerable difficulties are frequently encountered in removing unwanted reaction products. The reaction in Figure 4.3(c) is preferred where possible. Thirdly, by reaction with organochlorosilanes to give a siloxane bond. This is the method most often used in the preparation of chemically bonded silica gel packing materials as it gives a stable product using a straightforward reaction scheme, and has largely superceded the methods shown in Figure 4.3(a) and (b).

Table 4.1 Commonly used bonded phases

Functional Group	Structure	Principle Use
Octadecyl	$-(CH_2)_{17}-CH_3$	Reversed phase
Octyl	$-(CH_2)_7-CH_3$	Reversed phase
Propyl	$-C_3H_7$	Reversed phase
Phenyl	$-C_6H_5$	Reversed phase
Aminoalkyl	$-(CH_2)_n-NH_2$	Normal and reversed phase
Cyanopropyl	$-(CH_2)_3-CN$	Normal and reversed phase
Diol	$-CH(OH)-CH_2OH$	Normal phase
Sulphonic acid	$-(CH_2)_n-SO_3H$	Cation exchange
Quaternary amine	$-(CH_2)_n-N^+(CH_3)_3$	Anion exhange

By modifying the side chain (R) on the organochlorosilane a wide variety of bonded stationary phases may be prepared, e.g. using $C_{18}H_{37}SiCl_3$ ($R = C_{18}H_{37}$) as the derivatisation reagent will mean the resulting siloxane contains the $-Si-O-C_{18}H_{37}$ group. This is the ubiquitous octadecylsilane or ODS phase, the preferred phase in reversed phase HPLC. Table 4.1 illustrates some of the bonded phases which are commercially available.

In a reversed phase separation using an alkylsilane bonded phase, the retention of the solutes depends upon the carbon loading, generally expressed as a percentage. For an ODS bonded phase, the carbon loading may vary from approximately 10% to 20%, whereas for an octylsilane (C8) bonded phase the value is more typically 5% to 10%. This implies that as the chain length of the alkyl group (R) increases then the solute retention will also increase, assuming that the percentage of the silanols derivatised does not change. The main difference therefore between the octadecyl (C18), octyl (C8) and propyl (C3) packing materials is that changing from one to another will alter the capacity factors (highest for ODS, lowest for propyl) although the selectivity of the separation between two compounds, i.e. their relative retention may not change a great deal. Generally the highest selectivity values will be obtained on ODS, which is one reason why this phase is by far the most popular for use in reversed phase systems. It is often suggested that ODS should be used for solutes with moderate to high lipophilicity and short-chained alkylsilanes selected for more polar compounds. Where the analyte is too hydrophilic to be retained on ODS, resulting in inadequate resolution, the use of a C3 phase (for example) is often preferable. The rationale for this choice of stationary phase is that the retention does not rely solely on hydrophobic interactions but also on other secondary mechanisms (e.g. polar interactions with free silanol groups). The use of octylsilane may also be considered for extremely lipophilic molecules which are held strongly onto the non-polar surface of ODS leading to unacceptably long analysis times. In practice however, it is possible to carry out HPLC analysis for most compounds on ODS and even in those cases where an alternative is required, a more radical change is often preferred to using a shorter alkyl chain.

All the reactions in Figure 4.3 illustrate a one-to-one relationship between the silanol group and the derivatisation reagent, resulting in one R group for each silanol. This type of phase is referred to as monomeric. By using a trichloroalkylsilane rather than a monochloroalkylsilane as the reagent in Figure 4.3(c), the product resulting from a one-to-one reaction can further react with either a nearby silanol or another chloroalkylsilane, leading to the build up of several layers. This type of phase is termed polymeric and while the process is more difficult to control than one designed to give a monomeric coverage, acceptable reproducibility can be obtained, and polymeric ODS phases are available. Due to their extra carbon loading, retention is generally higher than on monomeric phases.

The phenyl group offers a difference in selectivity to ODS for many compounds in reversed phase HPLC, and has been used in the analysis of polycyclic aromatic hydrocarbons. For most compounds, retention on a phenyl column is similar to that obtained on a C8 packing material.

The aminoalkyl and the cyanopropyl bonded phases are commonly used in that they can be equally well used in either the normal phase or the reversed phase mode. The versatility of the cyanopropyl column was exploited in the development of an expert system for HPLC, allowing a single column to be used for all solutes.

In the reversed phase mode, both of these bonded phases are principally used for highly polar materials which have little retention on the highly non-polar ODS phase. This is similar in principle to the use of short alkyl chained bonded phase described earlier. One example of this use is the chromatographic separation of carbohydrates on amino bonded phases. It is also possible to exploit the lower hydrophobic nature of these packings to decrease the capacity factor for compounds which would be strongly retained on ODS and thereby to reduce the analysis time. In addition, changing from an alkyl bonded silica to a cyano bonded silica changes the retention mechanism and may lead to profound changes in selectivity. If it becomes necessary to select an alternative phase to ODS for a reversed phase HPLC method, then changing to cyanopropyl rather than to C8 for example, gives a greater opportunity to optimise the resolution due to the selectivity changes.

In normal phase HPLC, the cyano and amino columns, in conjunction with the diol bonded phase columns, are now often used in preference to unmodified silica or alumina. Whilst silica can demonstrate remarkable selectivity for the separation of closely related isomers (e.g. E/Z isomers are commonly separated on silica), some of the operational demands of silica limit its usefulness. The difficulty of manufacturing reproducible microcrystalline silica and the possible effect of trace amounts of water on the retention of solutes on silica have already been discussed. An additional problem is that silica equilibrates rather slowly with the mobile phase, hence changing from one analytical method to another may be time consuming if a change of mobile phase is involved. Perhaps more importantly, the use of gradient elution on silica is fraught with problems, e.g. variable retention times for the solutes and long equilibration times between runs. The use of bonded phases in normal HPLC essentially overcomes all of these problems, particularly with regard to equilibration. The water content of the mobile phase is still an important factor due to its very high eluting power in normal phase systems, but is considerably less noticeable than on silica.

In comparison to silica, all three of these bonded phases display less polar character and therefore, for any given mobile phase, retention times will be shorter than on silica. In general, the diol and the cyano phases are more polar and hence give more retention than the amino phase, but due to the considerable selectivity differences which are apparent for many solutes, accurate prediction is difficult. Care should be taken when using amino bonded phases as they are known to react with carbonyl compounds under some circumstances to give imines (Schiff's bases). This irreversible condensation will not only render the column unusable (or at least change its surface chemistry considerably) but also means it will be a very long time before the peak elutes.

By chemically bonding a permanently ionised (e.g. a quaternary amine) or an easily ionisable group onto silica, a phase suitable for ion-exchange chromatography may be prepared. Ion-exchange chromatography is mainly used for simple ions (e.g. Cl^-, NO_3^-, Na^+) or organic compounds which are very easily ionised (e.g. carboxylic acids) and relies on the electrostatic interaction between the charged stationary phase and the ion in solution to effect retention.

For steric reasons, not all the silanols on the silica will react with the derivatisation reagent during the chemical bonding process discussed earlier, and any unreacted silanol groups will cause tailing of polar compounds, especially bases. To overcome this problem, the bonded phase material is normally treated with trimethylchlorosilane which, because of its smaller molecular size will have better access to unreacted silanols and hence produce trimethyl silanes. This procedure is known as end-capping. It is important to realise that even after a column has been end-capped, there will still be some unreacted silanol groups remaining (perhaps as much as 30% of the total silanols). Some of these may then be able to interact with the solutes during the chromatographic separation. Many of the problems associated with the use of ODS phases for the analysis of bases may be traced to the presence of silanol groups.

Not all silanol groups are the same; some are acidic and others are basic, although it is generally considered that the free silanols (Figure 4.4)

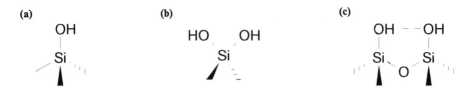

Figure 4.4 Silanol group types; (a) free silanol, (b) geminal silanol, (c) associated silanols.

are the most likely to cause problems in the chromatography of basic samples. The presence of metal ions on the silica surface may also influence the impact of a silanol group by affecting its acidity. The HPLC analysis of bases has been the subject of much discussion with many recommendations made for the best way to tackle the problem.

Some approaches concentrate on adjustments to the mobile phase, e.g. reducing the pH to approximately 4 or less to minimise the ionisation of the silanols. Another mobile phase modification is to add a basic compound, e.g. triethylamine or dimethyloctylamine (DMOA) to act as a blocking agent, thereby preventing the analyte from interaction with the silanol. These procedures are successful in many cases but are not universally applicable. Changing the pH of the mobile phase is not always an option as the solute may not be stable in acidic solutions and the pH of the packing material must also be considered. Chemically bonded silica will undergo hydrolysis of the siloxane bond and subsequent loss of the alkyl chain at pH values below approximately 2 and therefore it is prudent to keep the pH above 2.5. As mentioned earlier, silica is soluble in aqueous alkaline solutions and therefore pH values greater than 7.5 lead to rapid deterioration of the column. It is because of this fact that it is not normally possible to improve the chromatography by using a high mobile phase pH to suppress the ionisation of the basic solute. The use of silanol blockers is often successful, but reagents such as triethylamine are difficult to remove from the column, requiring extensive washing and resulting in long equilibration times when changing mobile phases.

Where changes to the mobile phase will not give a satisfactory result there remains the possibility of changing the stationary phase. Since the effect of the silanols depends on the type of silanol present, by selecting a silica with a low level of acidic silanols the unwanted interactions may be reduced. The acidic or basic property of the bonded phase is largely determined by the acidity of the silica used in its preparation, therefore if bases are to be analysed by reversed phase HPLC, it is logical to select a column packing with a low level of 'acidity'. Table 4.2 shows a number of commercially available packing materials ranked in order of acidity. The inference to be drawn from this table is that those columns close to the

Table 4.2 Ranking of some HPLC reversed phase columns in terms of 'acidity'

	Column
acidity increasing ↓	Zorbax RX
	Nucleosil
	μ Bondapak
	Partisil
	Spherisorb
	Lichrosorb
	Hypersil
	Zorbax
	Micropak

top of the list will be preferable for reversed phase HPLC of bases. It should be pointed out that batch variation may lead to insignificant differences between two adjacent packings in this table.

If it is correct to suggest that free silanol groups are mainly responsible for the acidic nature of silica gels, then producing a silica with a homogeneous distribution of silanols will encourage the formation of associated silanols (Figure 4.4) at the expense of free silanols. Such a silica was produced by Kohler & Kirkland (1987) and was then used to prepare an octyl packing material, now available from DuPont as Zorbax RX. As can be seen from Table 4.2 this product is the least acidic phase on the list even though the octyl bonded phase prepared by the same manufacturer using a more typical silica was the second most acidic.

The interest shown in this problem is such that many manufacturers have produced a special packing material designed to allow basic compounds to be chromatographed without the peak tailing and irreproducibility often encountered during HPLC analysis. The approach each manufacturer has taken may be slightly different, but in essence the focus has been on the reduction of acidic silanols on the silica gel surface. Many have concentrated on the elimination of trace metals, either by strict quality control of the process or by subsequent removal by acid washing. It was suggested that a shorter alkyl chain (e.g. C8) gives better results than ODS, presumably because of a more complete coverage of the silanols. Whichever methodology is adopted, the 'base deactivated' columns now available offer a much improved performance over traditional ODS phases for the chromatography of bases.

4.3 Polymer packings

A more certain method of ensuring that silanol based interactions do not spoil a chromatographic separation is to use a column which is not silica

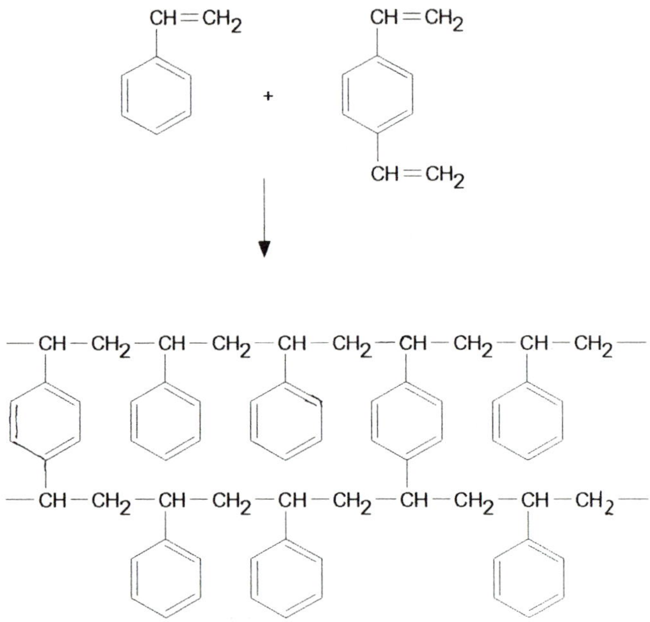

Figure 4.5 Synthesis of cross-linked polystyrene resin.

based. HPLC stationary phases are available which are built up from cross-linked poly(divinylbenzyl)styrene, prepared by polymerisation of mixtures of styrene and divinylbenzene (Figure 4.5).

These phases are used without further chemical modification, normally in the reversed phase mode with moderately polar mobile phases. Apart from the fact that silanol groups are absent, these packing materials differ in several ways from silica based types. One of the most useful differences is that they are stable over a wide pH range, commonly quoted as from 1 to 13. This allows a much greater choice of mobile phases than would be advisable with silica and, in the case of bases permits the ionisation to be suppressed by operating at high pH values.

When used in the reversed phase mode for HPLC, these phases give a similar or slightly greater retention than an ODS phase with a high carbon loading. Although there are clear advantages in using polymer based phases in preference to silica, their use has been rather limited for reversed phase HPLC, no doubt mainly due to the somewhat lower efficiencies available on these materials.

The cross-linking process used for the poly(divinylbenzyl)styrene phase produces rigid, spherical particles with a well-controlled pore size. This makes them ideal for use in size exclusion chromatography and indeed they are amongst the most important materials in use for this separation technique.

4.4   Porous graphitic carbon

Porous graphitic carbon (PGC) has a rigid structure, is chemically stable and has a very hydrophobic surface. It can therefore be used for reversed-phased HPLC, where it displays a similar or greater degree of retention to ODS. It has excellent pH stability and hence ion suppression of either acidic or basic drugs is easily carried out on this material. It displays different selectivity to silica based packings, allowing isomeric separations more typical of unmodified silica to be carried out (Bassler, 1989).

4.5   Solvents for HPLC

There are several criteria which a solvent should meet before being considered suitable for use as a mobile phase component HPLC. Firstly, a suitable solvent should be readily available in a pure form and have an acceptable level of toxicity. To keep the pressure drop across the column to a minimum, the solvent should also have a low viscosity. Consideration of the HPLC system as a whole involves attention being given to the solvent's compatibility with the detection method used as well as to its use as the mobile phase in a chromatographic separation. As the most commonly used detection method in HPLC is by ultraviolet (UV) radiation, the UV absorbance of a potential solvent is therefore often of extreme importance. Where refractive index detection is used, the best sensitivity will be obtained by maximising the difference between the refractive index of the mobile phase and that of the analyte, and hence the refractive index of a liquid may also be a parameter to be considered in solvent selection. Finally, the solvent must be able to dissolve the sample components completely without chemically reacting with them. Whilst no solvent can fulfil all the above criteria in all cases, the immense number of candidate liquids can at least be reduced to manageable proportions by applying the benchmarks listed above.

With the guidelines discussed in mind, consideration of the requirements for the mobile phase for an HPLC system can now be made. In contrast to gas chromatography, the mobile phase in HPLC is of primary importance in determining the quality of the chromatography. The design of a good chromatographic system is about achieving good resolution between the components of interest in a short time and detecting them with adequate sensitivity. Therefore, the aims of the separation part of the system can be broken down into attaining a reasonable capacity factor (k') for the solutes, a suitable selectivity (α) and a high efficiency (N). The relationship between each of these terms and the chromatographic resolution (Rs) is given by the following equation (chapter 2, equation (2.47))

$$Rs = \frac{1}{4}\sqrt{N}\frac{k'}{(k'+1)}(\alpha+1) \qquad (4.1)$$

As the solvents used will affect both the capacity factors and the selectivity of the system, the mobile phase is selected on its ability to elute the solutes in an acceptably short time with a reasonable resolution between each one. The choice of mobile phase components may also be influenced by the desire to suppress the ionisation of one or more of the analytes or to reduce peak tailing. These considerations, whilst very important are generally viewed separately from the factors which affect the k' and α values.

In both normal phase and reversed phase HPLC, the eluting power or solvent strength of the mobile phase is mainly determined by its polarity. Although most analysts have a good idea of what the term polarity implies and could rank most common solvents in order of their polarity, a more quantitative description would be very useful in chromatography.

One measure of polarity is the solubility parameter (δ) which was suggested by Hildebrand. This is defined as the square root of a solvent's vaporisation energy (ΔE) divided by its molar volume (V);

$$\delta^2 = \frac{\Delta E}{V} \qquad (4.2)$$

As the data in Table 4.3 show, the solubility parameter reflects how a chemist might rank these solvents in terms of polarity, e.g. water as the most polar (highest δ) and hexane as the least polar (lowest δ) but also one of the difficulties with this measurement of polarity is highlighted. The solubility parameter suggests that tetrahydrofuran (THF) and carbon tetrachloride are very similar even though carbon tetrachloride is immiscible with water whilst THF is miscible with water in all proportions. A similar comparison may be made between chloroform ($\delta = 19.1$, water-immiscible) and acetone ($\delta = 20.2$, water-miscible).

The reason for this apparent contradiction is that the vaporisation energy of a solvent is affected by all of the interactive forces in which the solvent can take part and hence the total solubility parameter may be

Table 4.3 Solubility parameters for solvents used in HPLC

Solvent	Solubility parameter (δ) ($J^{1/2} m^{-3/2} \times 10^{-3}$)
Hexane	14.9
Diethyl ether	15.5
Carbon tetrachloride	17.8
Tetrahydrofuran	19.1
Chloroform	19.1
Acetone	20.2
Acetonitrile	23.9
Methanol	29.4
Water	47.8

broken down into component parts representing the contributions made by dispersive forces, dipolar interactions and proton donating accepting tendencies. A significant portion of the intermolecular forces which make up the total solubility parameter for THF arises from dipolar and acid/base interactions. This is also true for acetone, giving these solvents a strong similarity to water in this respect. The cohesive forces in chloroform and in carbon tetrachloride are in contrast, largely dispersive in nature.

Describing the total solubility parameter in terms of the individual interaction types which occur gives a much better correlation with experimental data; however, the calculation of the individual effects is not so straightforward and a variety of methods are in use which unfortunately give a variety of values. This has somewhat reduced the applicability of the solubility parameter for chromatographers although it still may be used for calculating the overall polarity of solvent mixtures.

An alternative approach uses the polarity index, P' proposed by Snyder. This is based upon experimentally determined gas chromatographic retention of three test solvents on a large number of stationary phases. The test solvents selected are ethanol, 1,4-dioxane and nitromethane. As well as an overall polarity index (P'), three other parameters are calculated, x_e (a proton acceptor parameter), x_d (a proton donor parameter) and x_n (a strong dipole parameter).

The relative polarity of solvents in the polarity index scheme differs from that in the solubility parameter scheme. For example, using δ values methanol is more polar than acetonitrile whereas using Snyder's polarity index, acetonitrile is more polar than methanol. Practical experience suggests that methanol is more polar than acetonitrile and therefore supports the solubility parameter approach.

Table 4.4 Polarity parameters for solvents used in HPLC

Solvent	P^1	x_e	x_d	x_n	Group
Hexane	0.1	*	*	*	*
Diethyl ether	2.8	0.53	0.13	0.34	I
Methanol	5.1	0.48	0.22	0.31	II
Ethanol	4.3	0.52	0.19	0.29	II
1 – Propanol	4	0.54	0.19	0.27	II
Octanol	3.4	0.56	0.18	0.25	II
THF	4	0.38	0.2	0.42	III
Acetic acid	6	0.39	0.31	0.3	IV
Dichloromethane	3.1	0.29	0.18	0.53	V
Acetonitrile	5.8	0.31	0.27	0.42	VI
Acetone	5.1	0.35	0.23	0.42	VI
Toluene	2.4	0.25	0.28	0.47	VII
Chloroform	4.1	0.25	0.41	0.33	**
Water	10.2	0.37	0.37	0.25	VIII

*Irrelevant due to extremely low P^1 value
**Close to Group VII

Table 4.5 Members of groups I–VIII

Group	Members
I	Aliphatic ethers
II	Aliphatic alcohols
III	Pyridine derivatives, THF, sulphoxides
IV	Glycols, acetic acid
V	Dichloromethane
VI	Aliphatic esters and ketones, nitriles, dioxane
VII	Aromatic hydrocarbons, halogenated aromatics, aromatic ethers, nitro compunds
VIII	Water

The main advantage of the Snyder system is that it allows solvents to be grouped according to the type of interactions in which they take part. As Table 4.4 shows, solvents of the same chemical type (e.g. the four alcohols listed) have very similar values for the selectivity parameters (x_a, x_d and x_n) even though the P′ value is different for each one. Plotting the solvent's selectivity parameters on a triangular diagram yields a representation as shown in Figure 4.6. All of the common solvents fall into one of eight groups (last column of Table 4.4) with the exception of chloroform which falls close to both groups VII and VIII but is outside the circle of both. Solvents belonging to the same chemical class fall into the same group, and therefore even if a solvent is not listed above, its probable grouping in the Snyder classification system may be inferred from its principal functional group. Note, however, that a group may contain solvents of very different chemical types, e.g. both toluene and nitromethane are in Group VII.

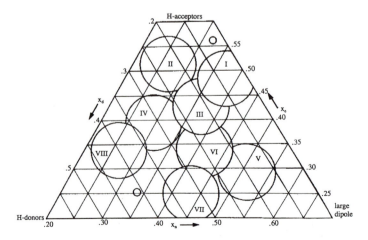

Figure 4.6 Selectivity triangle for HPLC solvents.

As solvents in the same group have the same or very similar selectivity values, if a chromatographic separation cannot be obtained with a particular solvent mixture then it is unlikely that changing one of the solvents for another in the same group will improve the separation. For example, changing a mobile phase from a hexane : ethyl acetate (Group VI) mixture to a hexane : acetonitrile (Group VI) mixture offers few prospects of improved selectivity; it would be preferable to use a hexane : THF (Group III) mixture or a hexane : propanol (Group II) mixture.

This suggests that instead of having a large number of candidate solvents, it is necessary only to select one solvent from each of the eight groups. The choice of solvent in each group may then be made on the basis of viscosity, UV absorbance, toxicity, etc. Another factor here may be miscibility, e.g. whilst propanol offers no selectivity advantage over methanol, it is miscible with solvents such as hexane which are commonly used in normal phase HPLC. With this in mind, it is logical to select methanol for reversed phase HPLC (low viscosity) and propanol for normal phase HPLC. The Snyder classification demonstrates that it is unnecessary to use solvents such as benzene, which is carcinogenic, as nitromethane would have the same selectivity properties.

When selecting the solvents to be used for a HPLC mobile phase the overall solvent strength is adjusted to give a suitable retention (k' value from 2 to 10) by mixing together one weakly eluting solvent and one strongly eluting solvent. The solvent with low eluting power which is used will naturally depend on the mode of HPLC in operation. In reversed phase HPLC, the weakest solvent and therefore the one most commonly used is water; in normal phase HPLC hexane or a similar hydrocarbon is the usual option.

The second solvent or modifier is then chosen from one of the remaining solvent groups. In reversed phase HPLC, this is most often one of three solvents; methanol (Group II), acetonitrile (Group VI) or THF (Group III). All three of these solvents have low viscosity, a reasonably low toxicity level and a low UV cutoff and most separations can be achieved with a mixture of one or more of them with water.

In normal phase HPLC, the range of solvents in common use is wider than in reversed phase HPLC, although a systematic approach would be to use one of either dichloromethane, propanol, methyl t-butyl ether or acetonitrile in conjunction with hexane. Other solvents such as chloroform, which has a slightly different selectivity or acetic acid, which may be selected for its strong proton donating powers are also useful. All of these solvents are compatible with UV detection and are relatively stable and therefore may be considered to be the preferred solvents in their groups.

A mobile phase of any given eluting power may be prepared by mixing two (or more) pure solvents together. Different mobile phases composed of different solvents but with the same overall polarity should give similar capacity factors for a set of solutes and such mobile phase mixtures are termed isoeluotropic, i.e. having the same eluting power. The capacity factors for each solute may vary slightly as the selectivity values of solvents used will differ and hence changing the mobile phase composition can improve the chromatographic resolution without adversely affecting the analysis time. This approach is widely used in selectivity optimization procedures (Chapters 3 and 7).

There are a number of ways of calculating overall solvent strength in order to give isoeluotropic mixtures. One method is to express the strength as the product of the sum of the strengths of the component solvents with their volume fractions, i.e.

$$S = \Sigma \, S_i V_i \qquad (4.3)$$

where S is the total solvent strength, S_i is the individual solvent strength and V_i is the volume fraction of the solvent.

The polarity index or the solubility parameter may be used as a measure of solvent strength, which would be a measure of polarity in those cases. For reversed phase HPLC, solvent strength parameters have been proposed for the four most common solvents used, i.e. water ($S_i = 0$), methanol ($S_i = 2.6$), acetonitrile ($S_i = 3.2$) and THF ($S_i = 4.5$). Using these values water makes no contribution to the eluting power of the mobile phase and the solvent strength is measured by the volume fraction of organic modifier.

For example, to calculate the percentage of acetonitrile (ACN) in a mobile phase which is isoeluotropic with a mobile phase containing 70% methanol (MeOH)

$$S = S_{MeOH} \times \%_{MeOH} = S_{ACN} \times \%_{ACN} \qquad (4.4)$$

$$2.6 \times 70 = 3.2 \times \%_{ACN} \qquad (4.5)$$

$$\%_{ACN} = \frac{(2.6 \times 70)}{3.2} = 56.9 \qquad (4.6)$$

Hence 56.9% acetonitrile is isoeluotropic with 70% methanol.

Another approach, based on a study of the retention of a wide variety of solutes by Schoenmakers, is to use transfer rules. This has been quite successful, perhaps because it is based on empirical data, although if the solutes in use are very different from the solutes used to develop the rules then their reliability may suffer.

$$\varphi_{ACN} = 0.32(\varphi_{MeOH})^2 + 0.57 \, \varphi_{MeOH} \qquad (4.7)$$

$$\varphi_{THF} = 0.66\varphi_{MeOH} \qquad (4.8)$$

where φ_{MeOH} is the volume fraction of methanol, φ_{ACN} is the volume fraction of acetonitrile and φ_{THF} is the volume fraction of THF.

5 Instrumentation: pumps, injectors and column design

T. NOCTOR

5.1 Introduction

The development of HPLC from open-column chromatography can be characterised by a trend to smaller and more regularly shaped stationary phase particles. As has already been explained elsewhere in this book, these factors, along with the narrowest possible particle size distribution, allow more efficient separations. However, the features of stationary phase particles which improve chromatographic efficiency also permit them to form more densely packed beds, and consequently to offer higher resistance to the flow of mobile phase. Thus larger demands are placed on the pumping system. The pumps formerly used in low pressure column chromatography proved to be generally unsuitable for HPLC; even if they could develop the torque required to drive mobile phase through the column, it was at the cost of large fluctuations in flow rate during the pumping cycle, severely reducing their usefulness. The development of HPLC has therefore had to be supported by a parallel development in pump technology.

The pump can be thought of as the heart of the HPLC system, the instrument that makes the mobile phase mobile. At first glance, the attributes required of the HPLC pump are simple: it should be able to deliver the mobile phase through the column at a constant and reproducible flow rate (or pressure), and be able to do this in a pulse-free manner. However, these things are not as easy to achieve as might first be imagined. While it is often a fairly straightforward proposition for a pump to attain the high pressures associated with modern HPLC, to do this without causing huge pulsations throughout the system is far more difficult. This is important because analysts increasingly need to detect (and measure) smaller and smaller amounts of substances, requiring very stable flow. Fluctuations in flow rate in the HPLC system can result in a 'noisy' chromatogram, in which small peaks can become 'lost'. This severely reduces the sensitivity of the method. In addition, modern HPLC pumps may be called on to operate, reproducibly and steadily, at very low or very high flow rates, where the performance specifications of older pumps would be less satisfactory. It is a testament to the ingenuity of the various manufacturers that modern HPLC pumps are routinely able to meet the very stringent requirements of analysts.

In the first part of this chapter, the different designs of LC pumps are dealt with, from the simple mechanical instruments which were used to pioneer the field, to the microprocessor-controlled pumps available today. Later, some other related topics are considered, such as the problems associated with the introduction of samples, and the means available for connecting up the highly pressurised HPLC system.

5.1.1   The ideal HPLC pump

In attempting to design an ideal pump for HPLC, the following features should be built into it.

5.1.1.1 Pressure generation.

It should clearly be able to generate the high pressures associated with modern HPLC columns, and cope with these over long periods of time. It should certainly have some means of indicating these pressures to the outside world. This facility allows the condition of the pump and the column to be monitored. Changes in pressure, whether intermittent or sustained, are usually a good indication that something is amiss with one or both of these components.

The pressure experienced by the pump as it forces mobile phase through the column is known as the column back pressure. This will vary of course, from column to column (according to dimensions and particle size) and with the flow rate and viscosity of the mobile phase used. The relationship between these parameters is expressed in the following equation:

$$P = \frac{L u \eta}{K^\circ \times 10^8} \tag{5.1}$$

where P is the column back pressure (bars), L is the column length (cm), u is the linear flow rate of the mobile phase through the column (cm/s), η is the viscosity of the mobile phase (cP) and K_0 is the permeability of the column bed (cm^2). In other words, the smaller the particle, the lower the permeability of the column bed, and so the higher the pressure. Similarly, increasing the length of the column, the flow rate or the viscosity of the mobile phase will all increase column back pressure.

In the above relationship, the back pressure is measured in bars. This is just one of a number of systems of units used in HPLC to measure pressure. The SI unit of pressure is the Pascal (Pa), although this is not used very often in this context, probably because of the large and unwieldy numbers needed to describe typical HPLC pressures. Other units that may be encountered include: pounds per square inch (psi), which tends to be the favoured unit in the United States, and its metric equivalent kg/cm^2.

For reference, 1 bar is equal to 1 atmosphere of pressure, which is approximately equal to 14 psi, 1 kg cm^{-2} or 10^5 Pa.

Back pressures in HPLC (for an average column 25 cm long, internal diameter of 4.6 mm, packed with 5 μm particles, with an aqueous mobile phase pumped at 1 ml min^{-1}) are of the order of 150 bar (2000 psi), although they can be much lower or higher (up to 300 bar) depending on the factors outlined above. Typical HPLC pumps sold today have maximum operating pressures around 500 bar (7000 psi), and so have a large degree of performance in hand during day-to-day operation.

5.1.1.2 Flow rate. The flow rate from the pump should be accurate, regardless of the system in which the pump is used. This means that the flow rate actually produced by the pump is the same as that dialled up on the front panel, and that this should not be affected by the rest of the HPLC system (for constant flow pumps, at least). The flow rate should be reproducible and practically free of pulsations. Pulsatile flow can limit the sensitivity of HPLC assays, resulting in a rhythmic variation in the apparent refractive index of the mobile phase flowing through the detector, which ultimately manifests itself in the chromatogram as baseline noise. Various means are employed to attempt to eliminate flow pulsation (see section 5.3).

The ideal pump should be able to provide a wide range of flow rates, while maintaining adequate levels of accuracy and precision. Flow rates in analytical HPLC vary from around $0.5–5 \text{ cm}^3 \text{ min}^{-1}$, with most applications using flow rates in the range $1–2 \text{ cm}^3 \text{ min}^{-1}$. Variants of HPLC use flow rates outside these ranges. Small bore HPLC (in which columns with internal diameters of less than 2 mm are used) requires very low flow rates, of around a few microlitres per minute. At the opposite end of the scale, semi-preparative separations (using columns with large internal diameters) require flow rates of up to $20 \text{ cm}^3 \text{ min}^{-1}$. The accuracy of the flow rate selected should ideally be less than $\pm 5\%$ of the nominal, and it should vary by less than $\pm 0.5\%$. Most modern HPLC pumps are able to perform within these specifications over a flow rate range of $0.01–10 \text{ cm}^3 \text{ min}^{-1}$.

5.1.1.3 Resistance. The pump should be inert to the various solvents, buffer salts and solutes to which it will be exposed in general use. In the vast majority of applications, stainless steel is used for metallic parts that contact the mobile phase, and this is perfectly acceptable. Elsewhere in the pump, resistant minerals such as sapphire or ruby are used for the pistons and in valve components subject to hard wear, and materials such as PTFE (Teflon®) are used for gaskets and seals.

The harsh eluents sometimes used in ion-exchange or hydrophobic interaction chromatography, which are used in the HPLC separation of biological molecules, may attack stainless steel. In such cases, all-titanium pump heads and pistons may be used. Titanium is sometimes used in such applications, as it is highly resistant to extremes of salinity and pH, and to attack from commonly used reagents such as formic acid. Pump heads constructed from titanium, or other inert materials, are essentially free from the problems associated with stainless steel systems, and are said to be 'bio-compatible'. They are, however, highly expensive.

5.1.1.4 Component reliability. The ideal pump should be mechanically reliable, but at the same time be designed so that components can be easily replaced. This is not as contradictory as it sounds. The pump is the most mechanically complex component of the majority of LC systems, and certainly has the most moving parts. Due to the stresses placed on the pump during operation, certain parts will inevitably wear out. With clever design, the parts that wear out will be cheap and easy to replace, and their regular replacement should prevent more expensive parts of the pump from wearing out. Components such as the seals that prevent gross leakage of mobile phase (piston seals), and constituents of the valves that allow unidirectional flow of the mobile phase (check valves), are therefore specifically designed to be replaced at regular intervals. Regular replacement of pump seals protects more expensive components of the pump from erosion, and so should be easy to carry out. Good pump design ensures that this is so.

A well-designed pump will allow easy access to the low-pressure side of the piston seal, to permit rinsing. It is here that slow leakage of buffer-based mobile phases leads to the formation of small salt crystals. Such crystals are highly abrasive, and greatly increase seal wear, as well as potentially causing scratching of the piston. This is a problem particularly when the pump has been left to stand idle for a while. On re-starting the pump, the sharp buffer crystals act as course-grade sandpaper on the piston and seal, rapidly causing potentially expensive damage. Regular washing away of crystalline build-up will greatly reduce this type of wear.

5.2   Types of HPLC pump

There are several different types of HPLC pump available. However, these all fall into two main classes; those which deliver mobile phase at a constant pressure, and those that pump at constant flow rates. In the vast majority (almost 90%) of current analytical HPLC work, it is the latter type of pump that is used. However, both types of pump have their place in modern chromatography, as discussed in the following sections.

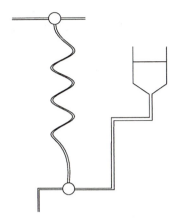

Figure 5.1 A diagrammatic representation of the pressurised coil pump.

5.2.1   Constant pressure pumps

Constant pressure pumps utilise pneumatics or hydraulics apply the pressure required to force the mobile phase through the column, either directly or indirectly. Two main designs of constant pressure pump exist; the pressurised coil pump, and the pneumatic pressure intensifier type. The pressurised coil pump is now all but redundant, but as it represents the most simple means possible of pumping at high pressure through an HPLC column it is described briefly.

5.2.1.1 The pressurised coil pump.   A diagrammatic representation of the pressurised coil pump is shown in Figure 5.1. A coil of tubing (capacity around 500 ml) is filled with mobile phase from a reservoir by means of a two-way valve. The valve is then switched so that it connects the coil to the head of the column, and a second valve at the top end of the coil actuated to allow pressurised inert, insoluble gas to enter. The pressure of the gas drives the mobile phase through the column. Pressures are limited to those available from bottled gas. This means that when operating at pressures above 100 bar, only about half of the tank can be used before pressure falls below that required to drive the mobile phase at the flow rate required. This results in heavy consumption of bottled gas, and, consequently, quite high operating costs. While the flow obtained from pressurised coil pumps is often relatively pulse-free, it is usually neither sufficiently accurate nor reproducible enough for the majority of modern requirements.

5.2.1.2 Pressure intensifier pumps.   The most common examples are pneumatic in principle, although hydraulic versions have been described.

They work by applying moderate gas pressure on a piston with a large surface area, which pushes forward a much smaller piston in contact with the mobile phase. The result of this is a large amplification in pressure, proportional to the area ratio of the two pistons. The effective pressures generated can be very high. Because of the pressure amplification factor, the pneumatic side of the pump can operate at fairly low pressure (about 10 bar) making the system economical to run, in terms of gas usage. At the start of the pump cycle, gas is introduced into the driving side of the large piston, forcing it and the high pressure piston forward and driving mobile phase through the column. At the end of the piston stroke, the gas is quickly transferred to the reverse side of the piston, forcing it rapidly back to its starting position, refilling the solvent chamber with mobile phase at the same time. The cycle is then set to recommence. Because of this backwards and forwards nature of the piston, single direction valves (check-valves) are required to ensure that flow is directed only to the column. As the return stroke is very quick, mobile phase is drawn very rapidly into the pump. Care should therefore be taken that any filter placed at the mobile phase inlet does not cause excessive resistance to flow, otherwise cavitation problems may occur (the sudden, strong negative pressure causes dissolved gases to come out of solution).

Other problems with pneumatic intensifier pumps include the fact that access to the high pressure seals for inspection and maintenance is usually quite restricted, by nature of their design. Because of the way they operate, the flow they produce is inherently highly pulsatile in nature and they also tend to be extremely noisy in use. For these reasons, pumps of this type are not used in general analytical HPLC. However, pneumatic intensifier pumps have found a niche in the packing of HPLC columns, where the intermittent nature of the function and their ability to deliver very high pressures compensate somewhat for their shortcomings in the analytical field.

5.2.2   Constant flow pumps

The vast majority of pumps currently used in analytical HPLC are of the constant flow variety. With this type of pump, changes in the chromatographic system, for instance those leading to variations in the back pressure experienced by the pump, are compensated for, and the flow rate remains constant; this is an important factor in most analytical applications. Two major types of constant flow pump exist: the reciprocating piston design and the syringe type.

5.2.2.1 Reciprocating piston pumps.   Around 85% of the HPLC pumps in current use are of the reciprocating piston type. Figure 5.2 shows a basic representation of their construction. With this type of pump, the mobile

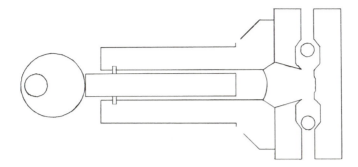

Figure 5.2 A basic representation of a reciprocating piston pump.

phase is forced through the chromatographic system by a piston (some-times called a plunger), which is driven by a powerful electric motor. The flow is in a single direction, due to the presence of the inlet and outlet check-valves. These are essentially one-way valves that allow mobile phase to enter the pump head from the reservoir, and to leave only in the direc-tion of the column. On the intake or backward stroke of the piston, nega-tive pressure opens the inlet valve, whilst simultaneously closing the outlet valve, allowing the mobile phase to fill the pump head. On the driving, or forward, stroke this situation is reversed; the positive pressure exerted opens the outlet check-valve, allowing the mobile phase to reach the column, while the inlet check valve is forced closed. Flow rate is modified by altering the frequency at which the cam driving the piston revolves.

As the piston is in direct contact with the mobile phase, it must be made from some type of material that is resistant to chemical attack (see above). For most purposes. pistons made from industrial grade ruby or sapphire are used.

A potential drawback of the reciprocating piston pump may by now be apparent. As only the forward stroke of the piston drives eluent through the column, then during the return stroke, flow through the chromato-graphic system ceases. This makes the flow from a single, unmodified reciprocating piston inherently pulsatile.

The original reciprocating pumps used eccentric circular cams to drive the piston. When a cam of this type is used, 50% of the pump cycle is given over to the filling of the pump head with mobile phase. The use of differently shaped (e.g. cardioid) cams allows the driving stroke to be greatly extended, and the filling stroke to be very much reduced. This sig-nificantly shortens the proportion of the pumping cycle during which the pump does not deliver mobile phase, but it does not eliminate completely the pulsatile nature of the flow. Three main strategies have been developed to overcome the problem of pulsation: the use of multiple piston heads,

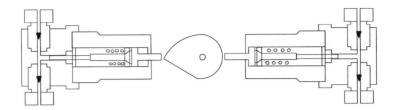

Figure 5.3 An arrangement to overcome the problem of pulsation using two reciprocating pistons, operated by the same motor but with their pumping cycles 180° out of phase.

the use of pulse dampening systems, and electronic control of piston speed.

Figure 5.3 shows an arrangement in which two reciprocating pistons are operated by the same motor, but due to the arrangement of their cams, their pumping cycles are 180° out of phase. Thus, when one piston is driving, the other piston is refilling, and *vice versa*. With clever cam design, the crossover between the driving phases of each piston can be extremely smooth, making the flow output from pumps of this type extremely stable. Dual piston pumps of this general type form the bulk of those in current analytical HPLC use. There was a brief trend towards pumps with three pistons during the late 1970s. Each of the three pistons was operated 120° out of phase, which resulted in an almost completely pulse-free delivery. However, the additional piston, seals and check-valves meant additional maintenance problems. HPLC pump technology in the meantime advanced to the point where equivalent performance could easily be obtained from a dual piston system, largely consigning the triple headed monster to the shelf of chromatographic history.

5.2.2.2 Syringe pumps. One of the first types of HPLC pump to be specifically built for the purpose was the syringe type. Such pumps are essentially very simple in design, employing a stepper motor to drive what is effectively the plunger of a large syringe to push mobile phase at a constant flow rate through the column. The capacity of the syringe can be virtually anything up to 500 ml. The flow rate from syringe pumps is extremely constant, and can be accurately and precisely controlled by altering the speed of the stepper motor.

Syringe pumps are not widely used, due to the tedious need with some models to dismantle the syringe in order to refill it with mobile phase. Other designs of syringe pump do not need to be dismantled for refilling, but use a check-valve arrangement to allow the reverse stroke of the plunger to automatically refill the syringe chamber. However, this process itself is very slow.

A further problem associated with syringe pumps arises from the fact that most of the fluids used as mobile phases in HPLC can be compressed, to some degree. This is especially the case with the purely organic solvents used in normal phase HPLC, some of which are reduced in volume by up to 0.015% for each bar of pressure that is applied. So for example, 200 ml of a mobile phase having a compressibility of 0.01% per bar, being driven at 100 bar to give a flow rate of $1 \, cm^3 \, min^{-1}$ will require at least 10 minutes to reach its nominal flow rate. The initial displacement of the plunger only serves to compress the mobile phase.

Syringe pumps are very well suited to very small-bore HPLC. Such separations require the very low flow rates and excellent precision provided by syringe pumps. Due to the very small amounts of mobile phase consumed, the syringe has to be refilled only infrequently, making that aspect of the use of syringe pumps much less of a disadvantage. With the growing interest in small, and micro-bore, LC it is likely that the use of syringe pumps will increase once more.

5.3   Approaches to the reduction of flow pulsation

As stated above, it is highly desirable that the flow generated by a pump should be practically free from pulsations. Pulsations in flowrate can reduce the sensitivity of a method, and in extreme cases, damage the column. The most commonly used class of analytical HPLC pump, the reciprocating piston type, inherently provides relatively constant flow – particularly when multiple pistons are used. However, the current drive towards the detection of smaller and smaller quantities means that levels of pulsation that were perfectly acceptable only a few years ago are now regarded as unsuitable for many applications. The minimisation or eradication of pump pulsation may be achieved by mechanical or electronic means, or by a combination of the two.

5.3.1   Mechanical or physical pulse damping

A number of physical devices have been developed to minimise flow pulsations. The most simple examples employ a reservoir of fluid between the pump and the injector, which absorbs some of the excess energy of the pulse. At its simplest, this may comprise a simple length of tubing placed in-line after the pump. Alternatively, air-filled snubbers may be used. These are made up of a length of tubing which is sealed at one end. On the driving stroke of the piston, the air in the tube is compressed. On the reverse stroke, the air expands again, the overall effect being a smoothing of the flow profile. A further system employs a diaphragm which divides the mobile phase from a chamber filled with another fluid. Pulses in the mobile phase are transferred to the fluid on the other side of the diaphragm, resulting in a smoother flow reaching the column.

The major disadvantage of physical damping systems is that they generally tie up relatively large volumes of mobile phase, sometimes in poorly accessible nooks and crannies. A change in mobile phase requires that considerable volumes are passed, before eluent of the desired composition reaches the column.

5.3.2   Electronic pulse compensation

Over recent years, the manufacturers of HPLC pumps have taken advantage of the dramatic increase in the power/cost ratio of microprocessors, to equip pumps with a semblance of intelligence. Several pumps use processor-controlled sensors to finely govern the flow profile. The sensor is located downstream from the pump head, and precisely monitors the flow rate passing over or through it. When a small change is detected, the sensor feeds back a message to the stepper motor driving the piston, telling it to change its speed appropriately. Small changes in flow are thus compensated for, and the flow rate from the pump remains within extremely tight tolerances.

5.4   The modern HPLC pump

A discernible trend in modern pump design is towards the use of very small pistons (stroke volume around 100 μl). These tiny pistons have to be operated at very high driving speeds, to provide the flow rates required in analytical HPLC. However, when they are used in combination with electronic feedback pulse control mechanisms, as described above, they can provide extremely stable solvent delivery characteristics. Such is the efficacy of this approach that single piston pumps designed in this way are able to easily out-perform older dual piston pumps, and are consequently beginning to account for a major part of the LC pump market.

Many manufacturers have used the capabilities of on-board processing power to provide a user friendly interface for the operator. This results in a host of useful (and some not so useful) features for the analyst, and allows the pump to be remotely controlled by a computer, or another component of the HPLC system. This can facilitate method development for instance, as the pump can be programmed to change automatically the composition of the mobile phase after a number of runs. It does this by changing the proportion of time that it draws from each of a number (up to four) of solvent reservoirs, that can contain different modifiers or buffers. Computerisation of pumps can also allow unattended runs to be performed with a greatly increased measure of confidence. In a networked system, failure of one of the components can trigger the shutdown of the whole system. If, for instance, the detector lamp fails, then the entire system can cease operation, saving unnecessary wastage of mobile phase and possible loss of valuable samples.

Many modern pumps have in-built electronic logs and diagnostic systems, which remind the operator when a routine maintenance procedure is due, or inform that a certain component is performing outside tolerance levels. The offending part may then be replaced before it has a chance to fail completely at some later, and much more inconvenient, time.

5.5 Valve injectors

In order to obtain maximum chromatographic efficiency, the sample to be analysed should be introduced onto the head of the column as an extremely narrow band. The high pressures associated with HPLC require that specialised measures are taken to introduce the sample into the chromatograph. Originally, injection onto the HPLC column was made through a septum, using a sharp needle. However, there were a number of serious drawbacks to this approach, including short septum life, and potential blockage of the column with the small pieces of septum inevitably dislodged by the needle. Today, the vast majority of HPLC systems use valve-type injectors. Several designs of valve injectors exist, but all have certain common features. For instance, all use a 'loop', which can be either internal or external to the valve, to control the volume of sample injected, and to hold it prior to introduction onto the column. In the case of valves having internal loops, the 'loop' is actually a small channel engraved on a Teflon disc, which can accommodate up to 10 µl of sample. External loops are lengths of tubing, of precise volume (between 2 µl and 20 ml) which can usually be changed to allow different volumes to be injected. In both cases, it is good practice to flush through approximately 10 times the nominal volume of the loop, to remove residual sample from previous injections, and to ensure maximum precision. The valve allows the loop either to be isolated from the stream of eluent from the pump ('load' position, Figure 5.4a), or to be positioned in it (the 'inject' position) (Figure 5.4b). The loop is filled with sample from a syringe when the valve is in the 'load' position. Excess sample is lost through a drain port. Turning the valve to the 'inject' position allows the sample to be flushed from the loop by the mobile phase coming from the pump, taking it onto the column for separation. The advantages of valve injectors include: excellent precision (when used as recommended); compatibility with the pressures encountered in HPLC; lack of septa and their associated problems; and good facility for automatic operation. There are, however,

(a)

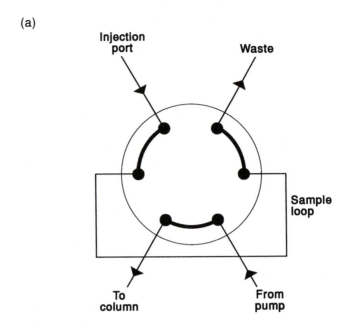

(b)

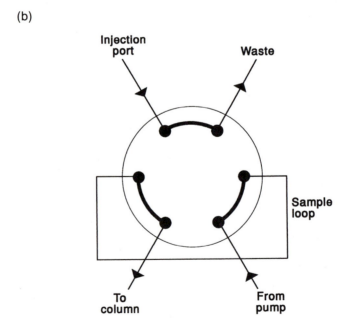

Figure 5.4 Valve-type injector. The valve allows the loop to be (a) isolated from the stream of eluent from the pump (load position) or (b) positioned in it (inject position).

several disadvantages associated with the use of valve injectors, although these are minor in comparison to the advantages of this approach. The valve is another part of the HPLC system that can contain seals, which if they fail, can cause the system to leak. The design of the valve injector involves the intimate meeting, under considerable pressure, of two discs: the stator and the rotor (it is the rotor that accommodates the channels indicated in Figure 5.4). Small particles from the sample may scratch either disc, causing a path by which mobile phase can leak. Valves can also be the source of artefact peaks in chromatograms. For instance, it is possible to contaminate one sample with material from a previous injection, if the loop is not well washed between the two, resulting in extra peaks in the second chromatogram. Finally, some valves require relatively large volumes of sample to work efficiently; this is particularly the case with external loop valves. As mentioned above, an important advantage associated with the loop valve design of injector is that it is easily adapted for automatic operation. The valve can be actuated, under computer control, by pneumatics or electric motors, and can be combined with some mechanism for the introduction of a range of sample solutions. As many HPLC assays are used in routine, repetitive applications, the benefits that this gives in terms of increased sample throughput are considerable. Consequently, many industrial laboratories use automated sample injectors, or autosamplers, which store and inject large numbers of samples, regardless of the presence of the analyst. Modern autosamplers can even act as simple robots, and perform simple dilution or derivatisation manipulations on samples prior to injection.

5.6 Column dimensions

The column is the hub of the HPLC system; the place where all the chromatographic action happens. However, its importance is often overlooked. This is generally because the column is a rather unimpressive spectacle compared to the other, much more obviously high-tech components of the HPLC system. However, despite its unprepossessing appearance, the column can have a very large influence on the overall chromatographic process. The most important feature of the column that influences results is its size. Generally, columns (which are usually made from stainless steel) are between about 2.5 and 25 cm in length, and have internal diameters in the order of a few millimetres.

The optimum length of the column required for a particular separation is dictated by the number of theoretical plates needed to give the desired resolution. If the column is too short, then clearly the column will not have enough 'resolving power' to achieve the separation and if it is too long, then analysis time is needlessly extended. The most common column lengths used in regular analytical HPLC are 10, 12.5, 15 and 25 cm, with 15 cm columns being perhaps the most popular.

Most analytical HPLC columns have internal diameters (i.d.) of around 5 mm, the majority being 4.6 mm. This diameter does not arise by chance. When a solute is injected onto the column, it will spread axially (i.e. in a direction perpendicular to that of flow) in the column bed. If it spreads far enough in this direction, it will reach the inner wall of the column. In this region, even the best packed columns will have some irregularities in packing, which can contribute to broadening of the chromatographic band of the solute, and therefore to a reduction in efficiency. This region of irregular packing can be reduced, primarily by using columns with polished inner surfaces, but it can never be entirely eliminated. Smaller particles will give a smaller irregular region at the column wall, but at the cost of higher operating back pressures. Chromatographic theory tells us that a solute injected as a point at the top of a 10 cm column, packed with 5 μm particles, will be a band with an axial diameter of about 2.5 mm at the end of the column. In a well-packed 5 mm i.d. column, the solute does not therefore reach the wall region, and efficiency should not be impaired.

Recently, there has been a revival in interest in the use of HPLC of columns with smaller internal diameters (i.e. 2 mm or less). Two major features of chromatography with small-bore columns are behind this trend: a significant reduction in the consumption of mobile phase, and an increase in the detector signal for a particular mass of solute (i.e. increased sensitivity).

It is the linear velocity of the mobile phase that influences a chromato-graphic separation, rather than the flow rate dialled up on the pump. Small columns require lower flow rates than large columns to achieve a given mobile phase velocity. For instance, if a regular 4.6 mm i.d. column is operated at a flow rate of $2 \, \text{cm}^3 \, \text{min}^{-1}$, there will be a linear mobile phase velocity through the column of $120 \, \text{mm} \, \text{min}^{-1}$. To achieve the same linear velocity in a column with an i.d. of 2 mm, a flow rate of only about $0.4 \, \text{cm}^3 \, \text{min}^{-1}$ is required. Thus, while the same separation can be achieved in both columns, the analysis using the small column consumes only 20% of the mobile phase required by the large column. The use of small-bore columns can therefore mean significant reductions in the use of toxic, expensive solvents, with obvious benefits to the health of analysts and the environment in general, and with considerable financial savings.

There is another important benefit in using smaller bore columns. Because of the smaller geometries involved, the same amount of solute injected onto a small-bore column will elute in a smaller volume than that of a larger column. This leads to a higher response at the detector for a particular mass, and therefore to a significant improvement in sensitivity. The major disadvantage associated with the use of miniaturised HPLC systems is the need for certain specialised components. Because the column volume is so small, extra-column volumes which would be inconsequential in regularly sized systems become significant enough to contribute to band broadening. The extra-column volume of the system must therefore be minimised to a point where the band broadening it causes ceases to adversely affect peak efficiency at the detector. This requires the use of specialised (i.e. expensive) injectors, detector flow cells and connections.

5.7 Column inlet/outlet

The column packing has to be retained in the column by some means. This is usually accomplished by the use of a porous disc of metal (known as a frit) at either end of the column. Most common frits will retain particles larger than a few micrometres, which means that while they prevent the packing material from escaping from the column, they will also trap particles larger than this size arriving at the head of the column in the mobile phase. It is inevitable that the frit at the column inlet will eventually become clogged with mobile phase-borne particles. For this reason frits are usually designed to be easily replaceable.

The end-fittings of the column come in a number of designs, but all play the same role: they provide a means via which the column is connected to the other components of the HPLC system. This simple role has to be achieved without introduction of unwanted dead volume into the system, to avoid unnecessary band broadening.

5.8 Connecting tubing and unions

In order to attain the best possible efficiency from the chromatographic system, it is necessary to exercise some care in the choice and use of the tubing used to connect all of the individual components. In particular, the tubing and any connections should not chemically interact with the solutes or mobile phases used, and should not contribute excessively to the dead volume in the system. In regular-bore HPLC, most analysts use precision engineered stainless steel tubing (which is often 'passivated' to reduce interaction with solutes) with an outer diameter of 1/16 inch, and various internal diameters.

Between the pump and the injector, solute diffusion is obviously not a problem, and so the i.d. of the tubing used here does not have to be particularly small. In fact, tubing with a slightly larger i.d. than generally used in other parts of the system is to be preferred, as this offers lower resistance to flow, and is less likely to be blocked by small bits of debris coming from the pump. However, there is a limit on the width of the inner diameter of the pre-injector tubing, because if the volume of tubing between the pump and the injector is too large, the length of time required for a new mobile phase composition to reach the column will be prolonged. This is particularly a problem in gradient work. For this reason, the tubing used before the injector often has a 'compromise' i.d. of around 1 mm.

After the injector, the smallest possible extra-column volume is required. This means that tubing with the narrowest possible i.d. should be used, in the shortest possible lengths. Commercial tubing is available with i.d. down to around 0.15 mm.

The use of steel tubing with very narrow i.d. has certain disadvantages. It is very easy to collapse the bore of the tube when bending it. This is a potential risk when connecting the column to the injector or to the detector which often involves creating some sharp angles in the tubing. Very fine bore tubing is easily blocked by small particles that may dislodge from various parts of the chromatographic system, or which result from improper mobile phase preparation. Another difficulty is that steel tubing is troublesome to cut effectively. Many obvious methods of cutting tubing result in either collapsing the tubing, or leaving unacceptably ragged ends (burring). Burred ends can be filed and polished, but this is a laborious process. Fortunately, special tools can be purchased to cut steel capillary tubing properly, and these should be used when available.

Many of the problems associated with the use of steel connecting tubing can be avoided by using tubing made from the flexible polymer, polyether etherketone (PEEK). PEEK is an inert, biocompatible polymer, which, when used to make tubing, is able to withstand the pressures encountered in HPLC, and yet is easily cut with a razor blade. However, as far as most analysts are concerned, the major advantage of PEEK is its great physical flexibility. The tubing may literally be tied in a knot without affecting the flow through it. In addition, because it is used with removable, finger-tight fittings, assembly and disassembly of the system is very easy. The main disadvantages of PEEK tubing are its incompatibility with certain solvents (e.g. tetrahydrofuran, methylene chloride and sulphuric acid), and its slightly higher cost than alternative tubing.

Connecting tubing is physically linked to the column-end or other fitting by a threaded nut. As there are a number of commercial types of column-end fitting available, then so there are a corresponding number of types of nut. It is important to use the correct type of nut for a particular end fitting, otherwise significant dead volumes may arise, due to an imperfect fit between the two. Generally, nuts are made from stainless steel. A seal is formed using a ferrule. This is a small, steel object, shaped something like a pear with a hole running down its centre. The ferrule slips snugly onto the connecting tube, between the nut and the end of the tube, with its wider end against the end of the nut. When the nut is tightened in the column-end fitting (or wherever a seal is required), the ferrule is compressed and deforms to grip the connecting tube very tightly, and to form a liquid-tight seal with the fitting. The connection between a steel ferrule and the connecting tubing is permanent, and so once a ferrule has been applied to a piece of tube, the only way it can be removed is to cut the tubing. Recently, nuts and ferrules, or combinations of the two, made from polymers have become popular. These are generally known as finger-tight fittings, because they can be used without the use of a spanner. They offer advantages in that they allow rapid changing of a column, or other system component, and they do not become permanently attached to a particular piece of tubing.

Whichever type of fitting is used, if the correct nut and ferrule are used, there will be almost no dead volume in the connection.

5.9 Conclusion

In summary, analysts expect a great deal of the component parts of their HPLC system, particularly from their humble HPLC pump. They expect it to work week in, week out, to high levels of tolerance, often without truly appreciating the technology and engineering required to satisfy their requirements.

6 Instrumentation: detectors and integrators

D.K. LLOYD

6.1 Ideal detectors and real detectors

The purpose of the detector in an HPLC system is to identify the presence of certain compounds of interest in the eluent from the HPLC column. Within the detector the analyte undergoes some physicochemical interaction, by which the analyte is recognised. This interaction is often optical in nature, examples being the measurement of the UV absorbance or fluorescence of the eluent. There are a number of other possibilities, such as electrochemical reactions, or the coupling of HPLC with a mass spectrometer for detection. These modes of detection have been used with all types of analytical chromatography right down to microcolumn techniques such as capillary electrophoresis. The detector provides an electrical output signal which is proportional to the quantity of the analyte within the detector at a given moment. This signal is then fed to an integrator, which is used for quantification of the amount of analyte present. Detectors can be classed into two broad categories, bulk property and solute property detectors. Bulk property detectors respond to any change in a certain physical property of the HPLC eluent, where that property is common to the mobile phase and analyte, and its magnitude is altered by the presence of the analyte. An example is the refractive index detector. Solute property detectors respond to certain unique properties of the analyte, such as fluorescence or electrochemical activity, with little or no contribution from the mobile phase.

There are a number of features which are generally desirable in HPLC detectors. In many instances, one of the most important requirements is that the detector is highly sensitive, thus capable of detecting very small amounts of analyte. A second property which can be very useful is the ability of a detector to selectively respond to certain compounds and not others. Related to selectivity is insensitivity to changes in the mobile phase, allowing the use of gradient elution. If the HPLC column gives complete resolution of a given compound from other sample constituents, a simple mixture of mobile phase plus this compound will pass through the detector. In such a situation, both a selective detector with response only to that compound, or a universal detector with response to any compound could be used to quantitate the analyte. If resolution is incomplete, mobile phase containing more than one dissolved component will pass through the detector. With a single detector which responds to both components, quantification of either analyte will be difficult, because of the resulting overlapping peaks. However a selective detector with response to only one component will still allow the quantification of that single compound.

The ultimate in selectivity in HPLC detection is seen with the use of mass-spectrometric detection, and for many applications this could be seen as the 'ideal' detection method. However, more mundane considerations such as size of the instrumentation and limited budgets combine to reduce HPLC-MS to a relatively small number of applications which most effectively exploit its unique properties. When such practical constraints are taken into account, the 'real' detector connected to the HPLC system usually turns out to be a device that is a compromise, and its performance characteristics need to be taken into account during the development of many analyses just as much as the performance of the column or any other component of the HPLC system. For example, lack of detection selectivity may require extra method development to completely resolve an interfering peak, or lack of sensitivity could force the inclusion of an extraction–concentration step in an analytical method to achieve detectable levels of analyte.

A listing of the most common types of HPLC detector is given in Table 6.1, along with some of their properties. By far the most widely used HPLC detector is the UV absorbance detector. This is due to a combination of factors: firstly, although sensitivity is not good compared to other detector types (e.g. fluorescence, electrochemical), sensitivity is adequate for the majority of HPLC analyses; secondly, a great many compounds are detectable by UV absorbance measurements; thirdly, simplicity of construction and economies of scale mean that UV absorbance detectors

Table 6.1 Properties of some detectors

Detector type	Sensitivity	Selectivity	Range of application	Characterise solute?	Gradient elution?	Relative cost
UV-vis	**	**	***	*	**	*
Spectrophotometric UV-vis	**	***	***	**	**	**
Fluorescence	***	***	*	**	**	*_**
RI	*	*	**	*	No	*_**
Electrochemical (amperometric)	***	***	*	**	*	*_**
Mass spectrometry	*_**	***	**	***	*_**	***

* low/poor; ** moderate; *** high/good.

are generally inexpensive compared to other types. In the majority of cases where a general-use HPLC system is purchased, a UV absorbance detector comes almost automatically as part of the package.

6.2   Detector performance criteria

The performance of all HPLC detectors can be characterised by certain parameters such as sensitivity, noise, drift, limit of detection, linear and dynamic range, and detection volume. Other factors are more specific to individual types of detectors, and are discussed in their respective sections.

6.2.1   Noise and drift

An 'ideal' detector might be expected to output an electrical signal which remained constant (e.g. at zero volts) throughout the chromatographic analysis except when an analyte entered the detection volume, whereupon the output signal would vary as a linear function of the analyte concentration within the detector volume. Reality is not quite so simple, with two added quantities, noise and drift being superimposed onto the detector output signal. Noise is a random short-term variation in the output signal. Variations occurring on a sub-second timescale, much more rapidly than variations in response due to elution of an analyte peak, are referred

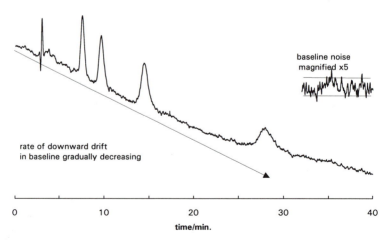

Figure 6.1 Example chromatogram illustrating noise (short-term fluctuations, approximate peak-to-peak value shown by lines on expanded section) and drift (the slope in the baseline, indicated by the arrow). The signal-to-noise ratio for the first three peaks should be sufficient for reasonably adequate quantitation, but the broader fourth peak on the noisy, sloping baseline would pose problems for most integrators.

to as short-term noise. Short-term noise gives the baseline a fuzzy appearance, and although it can lead to some difficulties with integration, it does not seriously interfere with the identification of a given peak. Long-term noise, comprising random variations in the output signal with a timescale similar to peaks due to the elution of compounds from the HPLC column can lead to confusion as to the identity of a peak. Drift is a very slow variation in the detector output signal, occurring on a longer timescale of minutes or hours. Noise and drift are illustrated on the chromatogram shown in Figure 6.1. Both noise and drift are undesirable. Noise limits the size of signal (and thus the amount of analyte) which can be reliably quantitated. Drift may make integration difficult, and excessive drift outside the operating range of the detector or integrator will lead to loss of data.

Noise arises from a variety of sources. The most basic is shot noise, arising in electrical circuits because of the discrete (quantum) nature of electrical current. Shot noise is an example of a white noise source, with an equal distribution of noise at all frequencies. Shot noise is found in all optical and electrical devices. Another important source of noise is flicker noise, a complex and wide ranging phenomenon. The characteristic property of flicker noise is that it is greatest at low frequencies, and there is a reduction in flicker noise amplitude with an increase in the frequency of measurement, f. Flicker noise has a power spectrum proportional to f^{-a}, where $a \approx 1$, generally. If shot and flicker noise are present together (this is usually the case), flicker noise will dominate at low frequencies, but above a certain cut-off frequency, shot noise will dominate. Devices such as lamps and lasers are particularly prone to flicker noise, and this can be a problem since considerable flicker noise may occur on a chromatographically significant timescale. Many detectors use dual-beam optical designs to cancel out the effects of flicker noise. Electronic filtering is provided to give smoothing of the output signal to reduce the effect of short-term noise. Usually there is a degree of operator control available over the degree of smoothing, provided by a rise-time or time constant control. Increasing the time constant increases smoothing, reducing short-term noise. Too long a time constant will result in the distortion of the peak shape of rapidly eluting peaks; as a rule of thumb, the time constant should be at most approximately one-tenth of the peak width time for a peak with $k' \approx 1$.

As well as noise arising due to particular electrical or optical devices, noise may be caused by external sources, such as mechanical vibrations, or pick-up of external electrical signals. Detector noise tends to increase with the age of the instrumentation. Certain components have limited lifetimes and are considered as expendable. An example is the deuterium lamp of a UV absorbance detector, which typically has a rated life of 2000 h or so. After this time the lamp will probably not fail completely, but the total light output will have slowly decreased to perhaps half of the output when the lamp was new. This often leads to higher detector noise levels. Other electrical components may degrade much more slowly, over a period of several years. Simple matters of cleanliness of the flow cell can cause increased noise levels in all types of detectors, and an observed increase in noise can often be remedied by a simple cleaning procedure. Sometimes noise and drift are observed to be a function of the mobile phase flow rate; in some cases noise peaks may be seen which correspond to the pulsations of the pump pistons. Such noise cannot be cured in bulk property detectors, except by better control of the mobile phase flow and composition, and reduction of pump pulsing. With solute-specific detectors, flow-dependent problems usually lie in the flow cell, and cleaning adjustment or replacement of the windows will often effect a cure.

6.2.2   Sensitivity and limit of detection

The terms sensitivity and limit of detection (LOD) are often used more or less synonymously, with reference to the minimum amount of an analyte which the detector can usefully detect without the analyte signal being swamped by noise. However, the term sensitivity is also used to express the change in output signal as a function of change in analyte concentration, and so it is perhaps better to reserve sensitivity for this usage. LOD is defined in a number of different ways, and comparison of detector LOD specifications can often be confusing, and probably meaningless if the procedures for determination of the LOD are not spelled out exactly. The performance of a detector is often specified in terms of the physical property which is being measured, thus the specification of a refractive-index detector would be given in terms of a detector noise level of so many refractive-index units. Perhaps the only way to really compare the LOD for two different detectors is to try them side by side on the same HPLC system, preferably with a variety of analytes under a range of conditions.

A wide variety of definitions of LOD have been put forward. One of the most commonly used, and simplest to apply, is the definition that the LOD corresponds to a concentration or mass of analyte (depending on which of these factors the detector response is proportional to) which produces a signal twice or three times the height of the peak-to-peak baseline noise. The peak-to-peak noise is illustrated in Figure 6.1. This is a fine working definition if noise is mainly short-term in character, but if there is a major component of long-term noise with a similar period to the signal peaks a higher (i.e. poorer) LOD may be more realistic; for example, claiming a LOD three or more times the peak-to-peak noise level. Another method of calculating LOD uses the regression line of detector response on sample concentration, which is calculated when making a calibration curve of detector response over a range of concentrations. The LOD is given by

$$\mathrm{LOD} = \frac{3\,s_{y/x}}{b} \qquad (6.1)$$

where b is the slope of the regression line, and $s_{y/x}$ is calculated from the y-residuals,

$$s_{y/x} = \left\{ \frac{\Sigma(y_i - y'_i)^2}{n-2} \right\}^{1/2} \qquad (6.2)$$

where y' values are calculated from the regression line, y values represent the actual measured detector response, and n is the number of data points.

The quantitation of analytes at concentrations close to the limit of detection can result in quite large errors, because of contributions to the measurement from noise. Sometimes these errors can be so large that a useful limit of quantitation may be defined, which is typically three times or more the limit of detection. This is rather dependent on the analytical precision that is required.

6.2.3   Linear and dynamic range

The linear range of a detector is simply defined as the range of analyte concentrations over which the detector output signal varies in linear proportion to the analyte concentration. After the maximum concentration within the linear range, an increase in sample concentration will probably still cause an increase in output signal, but without a linear proportionality. The region where a change in analyte concentration produces a change in output signal is referred to as the dynamic range, although without this necessarily being a linear relationship. For quantitative purposes, operation within the linear range is highly preferable.

6.2.4   Band-broadening, detector flow cells and time constant

In many HPLC detectors the column eluent flows through a cell within which some physicochemical interaction with the solutes takes place. Exceptions to this include mass spectrometric detectors where the eluent has to be vaporised before introduction into the vacuum system, or the evaporative mass detector, where again the eluent is heated and vaporised before undergoing analysis by light scattering. Often a number of flow cell options are offered by detector manufacturers, and this reflects the effect of the detection volume on the detected peak. The total peak variance, σ^2, is the sum of all the variance contributions,

$$\sigma^2 = \sigma_{lc}^2 + \sigma_d^2 \qquad (6.3)$$

where $\sigma_d{}^2$ represents variance contributions from the detector, and $\sigma^2{}_{lc}$ represents variance contributions from all other parts of the HPLC system. The two main detector parameters which contribute to band broadening are the flow cell volume, and the detector time constant. If the flow cell volume is large compared to the volume of the eluted HPLC peaks, considerable peak broadening will occur. Broadening will also be seen if the time constant is set at too large a value.

6.3   The UV-visible absorbance detector

6.3.1   *Principle of operation*

Many compounds in solution absorb visible or UV light, and this property forms the basis of the most popular detection system for HPLC. Absorption occurs when a photon of light interacts with a molecule to promote an electron from a lower to higher energy bound state. In the liquid state, electronic absorbtion spectra are generally relatively broad and featureless. Different mobile phases may be observed to cause variations in the absorbance strength and shifts in the wavelength of absorbance maxima. Some compounds have significant absorbances in the visible region of the spectrum (wavelengths of approximately 400–700 nm), whilst a great many more have strong absorbances at wavelengths in the ultraviolet region (below 400 nm). Certain very common structural features have strong absorbances at short UV wavelengths. Unfortunately, compounds with absorbances below about 220 nm, and particularly below 200 nm, are often difficult to detect in HPLC because of an overwhelmingly large background absorption due to the mobile phase.

The absorbance, A, of a solution is given by the Beer–Lambert law

$$A = \log\left[\frac{I_0}{I}\right] = \varepsilon\, lC \tag{6.4}$$

where I_0 is the intensity of light incident on the sample, and I is the transmitted light intensity. Alternatively, as shown above A can also be expressed in terms of the molar sample concentration, C, the pathlength of sample through which the light travels, l, and the molar extinction coefficient, ε. UV absorbance detectors provide an output signal proportional to absorbance rather than transmitted light intensity, which is useful since absorbance is directly proportional to the concentration of sample.

6.3.2   *Design of UV-visible absorbance detectors*

A block diagram of a UV absorbance detector is shown in Figure 6.2. Signal processing electronics ratio the signals from the sample and refer-

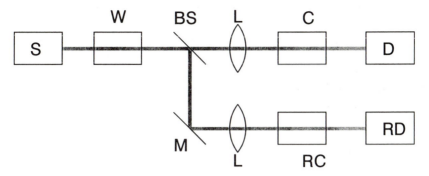

Figure 6.2 Block diagram of a UV-visible absorbance detector. Light from a source, S, is focused into a flow cell, C, through which the HPLC eluent is passed, and the transmitted light intensity is measured at a photodetector, D. A wavelength selection element, W, is placed after the source. A second light path is produced by the beamsplitter, BS, passing through a reference flow cell, RC, to the reference photodetector, RD. L are lenses; M is a mirror.

ence photodetectors, and this procedure corrects for drift and flicker noise in the lamp intensity, which would otherwise cause variations in the detector output. The processing electronics also serve to amplify the small signal from the photodetector into a signal suitable for connection to an integrator or chart recorder, and to give filtering of short-term noise. The degree of filtering is set by the time constant.

UV absorbance detectors fall into three main types, fixed wavelength, variable wavelength and spectrophotometric detectors, all of which have features in common with the basic instrument shown in Figure 6.2. The most simple design operates at a single, fixed wavelength. This type of design often uses a low pressure arc-lamp source with a line emission spectrum. Typical sources are the mercury lamp, with a strong emission line at 254 nm, and the zinc lamp with emission at 214 nm. A simple colour filter is used for wavelength selection to allow only light from the desired emission line to pass through the cell. A related variant on this type of design is the filter wheel detector, where several different filters can be rotated in front of the cell, to select different measurement wavelengths. In such a case, a deuterium lamp offering a continuous emission from around 190 nm to 600 nm can be used, allowing wavelength selection without the constraints of matching a limited number of line emissions.

Fixed wavelength and filter wheel detectors were very common in the 1970s and early 1980s, but today, few manufacturers of HPLC equipment have such devices in their product lines. Instead, variable wavelength UV and UV-visible absorbance detectors have become common. A modern, high-performance variable wavelength UV-visible absorbance detector may use a deuterium lamp for UV wavelengths and a tungsten lamp for the visible giving optimum light throughput and best sensitivity in each region of the spectrum. Wavelength selection is provided by a mono-chromator, a device which acts rather like a prism in causing a variable angular deflection of light depending on its wavelength. An exit slit in the monochromator allows light from a limited range of wavelengths (typi-cally in a band 1–2 nm wide, centred around the chosen wavelength) to pass through, and the selected wavelength is changed by rotating the monochromator with respect to the input light. Fibre optic devices rather than bulk optics may be used to reduce the number of optical surfaces in the system.

6.3.3 Operation of UV-visible absorbance detectors

The user must set a few operating parameters to use the detector. The main controls are generally the operating wavelength, the absorbance range, an autozero and the detector rise time. Setting the rise time has already been discussed. The range switch usually only affects the output signal at the recorder terminal. This output is usually 0–10 mV, with 10 mV corresponding to full scale on the selected range, and is used when connecting the detector to a strip-chart recorder. Most detectors have a second output for connection to a computer or integrator. This is typically 0–1 V, corresponding to 0–1 or 0–2 absorbance units (a.u.), and the output voltage is unaffected by the range switch. In this case, adjustment

Table 6.2 Approximate UV cutoff wavelength and RI of commonly used solvents

Solvent	UV cutoff (nm)	RI
Acetone	330	1.359
Acetonitrile	200	1.344
Benzene	280	1.501
Carbon disulphide	380	1.626
Carbon tetrachloride	265	1.466
Chloroform	245	1.443
Cyclohexane	210	1.427
Diethyl ether	220	1.353
Dimethyl sulphoxide	270	1.477
Ethanol	210	1.361
Ethyl acetate	255	1.370
Hexane	200	1.375
Methanol	210	1.329
Pentane	200	1.358
1-Propanol	210	1.385
Tetrahydrofuran	215	1.408
Toluene	285	1.496
Water		1.333

of the size of the signal appearing on the computer or integrator output is adjusted at the integrator itself. The autozero control is used to automatically adjust the detector output to 0 V. This is particularly useful when a drifting baseline threatens to take the signal out of the integrator or recorder range.

The wavelength must be selected to correspond to a chromophore of the compound(s) of interest. Having the UV spectra of the analytes is helpful, preferably measured in the same solvent as the HPLC mobile phase, and at a similar pH. The availability of a wavelength scanning HPLC detector is very useful for this. Alternatively, reference to handbooks such as the Merck Index often provides useful information. Many compounds have absorbance maxima with highest absorbance coefficients below 220 nm, but a longer wavelength absorbance with a weaker chromophore may give better results. This is because of reduced likelihood of interferences and reduced absorbance due to the mobile phase. Certain solvents are suitable for use at low wavelengths, for example acetonitrile–water mixtures may be used at 200 nm, but any traces of absorbing contaminants may cause high background absorbances and poorer sensitivity. Use of recirculated solvents may also be problematic. Table 6.2 shows the minimum useful wavelength for a variety of commonly used solvents. The problem of choice of solvent can be particularly acute when using gradient elution.

An operating wavelength chosen to correspond to an absorbance maximum is optimum not only because of increased sensitivity, but also because there is a relatively small change in absorbance with wavelength close to the maxima. If detection is made at a wavelength where the analyte's absorption coefficient varies rapidly as a function of wavelength, then small variations in the selected wavelength between analytical runs could lead to significant errors in quantitation of the solute. This is not usually a problem unless a change in wavelength followed by readjustment to the original wavelength has taken place. Typically, modern variable wavelength UV detector specifications allow ± 1 nm for wavelength accuracy, and ± 0.5 nm for wavelength reproducibility when returning to any given wavelength setting.

6.3.4 Spectrophotometric detectors

Over the last decade a number of manufacturers have introduced spectrophotometric detectors for HPLC, capable of producing real-time absorbance spectra of the column eluent. To be useful in HPLC the detector has to produce a number of spectra every second, so that even fast-eluting peaks can have a number of spectral snapshots taken. Spectrophotometric detectors offer advantages over conventional variable wavelength detectors, particularly for method development. With an

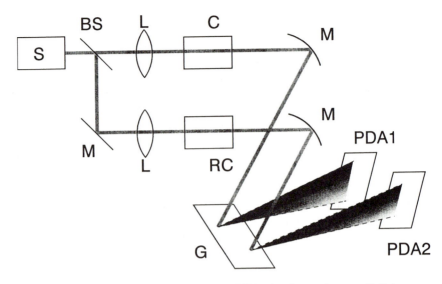

Figure 6.3 Block diagram of a diode-array UV-visible absorbance detector. S, light source; BS, beamsplitter; L, lenses; C, flow cell; M, mirrors; G, diffraction grating; PDA1, PDA2, photodiode arrays; RC, reference flow cell.

unknown sample, the spectrophotometric detector can be used for scanning over a wide wavelength range to be confident of detecting all the sample components. The spectral information obtained for each peak can be used along with k' for peak identification. By comparing the spectra obtained at the beginning and end of a peak, information on the peak purity can be gained. If a single peak has different spectra at the beginning and end of its elution, it is certainly comprised of more than one component.

Obtaining spectral information requires a method for either rapidly changing the wavelength of light focused through the flow cell to scan a spectrum in a fraction of a second, or transmitting a range of wavelengths through the sample and then splitting these with a holographic diffraction grating onto a series of detectors. The latter option is illustrated in Figure 6.3. A linear array of electronically scanned photodiodes is used, the output of each individual photodiode being integrated over a short period of time after which the total light signal is read out. In the design shown in Figure 6.3, spectra are taken in both the sample and reference channels. Less sophisticated designs may split a portion of the light from the lamp before the flow cell, providing the reference signal using the total light output from the lamp. One potential problem with placing the flow cell before the grating is that changes in the refractive index of the mobile phase may cause deflections of the light beam, which could lead to increased noise or drift. A second type of spectrophotometric detector uses a rapidly oscillating grating, mounted before the cell, to quickly scan a range of wavelengths. The design is essentially the same as the detector shown in Figure 6.2, but a low mass grating is mounted on an oscillating platform to allow scanning of a range of wavelengths. Whichever method of obtaining spectral information is used, a considerable degree of computing power is needed to manipulate and store the large amount of data that are generated by these instruments; this might be provided in the detector, but more frequently data handling occurs in a dedicated computer supplied with the detector.

6.3.5   Troubleshooting

A loss in sensitivity or increase in baseline noise in a UV absorbance detector is very often related to the HPLC mobile phase flowing through the detector flow cell. Rather less frequently, difficulties occur elsewhere in the detector optical system and occasionally in the detector electronics. The first step is generally to determine whether the problem occurs at the mobile phase and flow cell, or elsewhere. The most obvious sign of a flow-related problem is periodic noise on the baseline occurring in time with the piston strokes of the pump. This may be due to the presence of an air bubble in the flow cell which expands and contracts in response to the normal slight pressure fluctuations in the system, or an air bubble in the pump, causing excessive pressure variations. An air bubble in the flow cell is often visible if the cell is removed from the detector. In some detectors the operator can observe the flow cell without removal from the instrument. This should never be done while the UV lamp is operating because of the risk of eye damage from the UV light. Bubbles in the flow cell can usually be removed by intermittently increasing the back-pressure in the flow cell and then releasing it, by blocking the outlet tubing for a second or so while applying a mobile phase flow rate which is normal for the system, and then opening it again. This should be done with a degree of caution, since excessive back-pressure can easily damage the flow cell. Alternatively, a pump or syringe can be connected directly to the detector inlet, and liquid flushed through at a high flow rate. Any detector or flow restrictor after the detector under repair should be disconnected to avoid damage to the cell caused by overpressure. If the detector flow cell seems free of any bubbles, bleeding air from the pump may solve the problem. Faulty check valves can also give similar symptoms. If in doubt, exchange the pump (if possible) to see if pulsation still occurs. If the pulsation does not seem to be related to air in the system or a faulty pump, loose flow cell windows or poor flow cell alignment are likely causes. The presence of bubbles in the mobile phase which do not lodge in the cell often leads to noise spikes on the baseline. The cure in this case is to improve solvent degassing; if this alone is insufficient, increasing back-pressure at the detector using a flow restrictor such as a 5 m length of narrow-bore HPLC tube on the detector outlet may help. Again, care should be taken not to cause too high a back-pressure, which may damage the cell.

Many UV absorbance detectors have the facility to test the light intensity passing through the sample and reference channels, which can be useful in pinpointing a noise problem when there is no relation with flow of the mobile phase. A low reference and sample light level is usually indicative of an aging lamp which requires changing. Note that light levels should be measured at the wavelength specified by the manufacturer, since all lamps have a wavelength-dependent output intensity. In the case of deuterium lamps, a gradual deterioration occurs with age, and 2000 h of useful operating life is typical. However, some deterioration can occur just on storage of new deuterium lamps, and so keeping a large stock in reserve if they are not frequently used is not usually a good idea. Low light levels only in the sample channel can be due to a number of problems, including dirt on the flow cell windows, misalignment of the flow cell, a bubble in the flow cell or a highly absorbing mobile phase. When using microcolumn techniques such as capillary electrophoresis, it is frequent practice to use very low detection wavelengths, 190 nm or even less, to take advantage of the higher absorption coefficients of many compounds at these wavelengths. Operation is still possible since the short detection pathlength still allows enough light to reach the photodetector for useful measurements to be made. However, this is not a good reason to ignore the absorbance of the mobile phase, and improved results will still be achieved if this is kept to a minimum. Lot-to-lot variations in absorbing impurities in certain solvents and additives should be considered if a sudden increase in noise and drift is seen. Contamination of the photodetector, or a problem with the photodetector electronics is also a possibility. To clean the flow cell *in situ* the cell can be flushed for a few moments with a sequence of solvents, e.g. methanol, water, 20% nitric acid, water, then methanol. If this is not effective, flow cell removal and gentle cleaning of the windows using solvent and lens tissue, or even window replacement may be necessary.

6.4 The fluorescence detector

6.4.1 Principle of operation

When light is absorbed by a molecule, and an electron is promoted to a higher energy state, there are a number of pathways by which this energy can be dissipated, allowing a return to the ground state. Most commonly, energy is lost by transfer to surrounding molecules and contributes only to an overall heating of the environment. However, some molecules may lose only a part of their energy in this way, reverting to the lowest vibrational state of the electronic excited state. The return to the electronic ground state then occurs with energy being lost by emission of a photon, this process being called fluorescence. Because some energy is lost to general heating in the return to the vibrational ground state, the emitted photon has a lower energy and thus longer wavelength than the excitation light. Only a relatively small percentage of molecules lose a significant proportion of their excitation energy by fluorescence, and thus give strong fluorescence signals. For HPLC detection, it may be possible to structurally modify non-fluorescent compounds to make them strongly fluorescent.

6.4.2   Design of a fluorescence detector

The main attraction of fluorescence detection for HPLC is that for strongly fluorescent molecules, it can offer limits of detection two or three orders of magnitude lower than UV absorbance methods. The reason for this lies in the difference in the nature of the measurements. In a UV absorbance detector the photodetector is constantly illuminated at a high

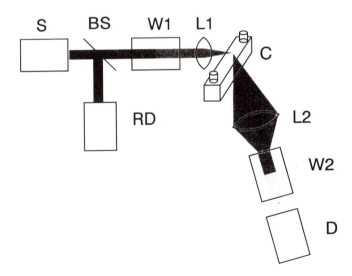

Figure 6.4 Block diagram of a fluorescence detector. Light from a source S is focussed through an excitation wavelength selection device, W1, and into a flow cell, C. Lens L2 collects fluorescent emission from the flow cell, and the collected light passes through the emission wavelength selection element, W2. The light then passes to a photodetector, D, the signal from which is passed to an amplifier and smoothing circuitry. A second light path is produced by the beamsplitter BS, passing to the reference photodiode, RD.

light level, and the presence of an absorbing compound causes a reduction in the light intensity. Thus at high sensitivity, the measurement is one of a very small change in a large signal. In a fluorescence detector, in the ideal situation the background light signal is zero, and light only appears at the photodetector when a fluorescent compound is present. In practice it is difficult to reduce the background signal below the noise levels of modern photodetectors, and in trying to achieve this goal lies much of the art of fluorescence detector design.

The block diagram of a fluorescence detector for HPLC is shown in Figure 6.4. The source is commonly a broad spectrum lamp such as a deuterium lamp, or a xenon lamp. Occasionally high intensity line sources such as mercury (254 nm) or zinc (214 nm) lamps are used, if their output maxima match the absorbance of an analyte of interest. Highest sensitivities can be achieved with intense laser sources; the disadvantage is the need to find a suitable type of laser with an output which matches the fluorophore of interest. When using a continuous excitation source the wavelength selection devices might be monochromators or filters, or one of each. With line sources the excitation wavelength selection is often by a matching filter. The flow cell is rather different in design than that of a UV absorbance detector; the requirement is to effectively illuminate the cell volume with the excitation light, while collecting the maximum possible fluorescent emission from that same volume and minimising the amount of extraneous light arriving at the photodetector. This extraneous light comprises scattered light from the cell walls, and possibly Rayleigh and Raman scattering from the cell and flow stream. A square-channel flow cell with excitation and emission orthogonal to each other is the most favourable arrangement for minimising scattered light. However, such favourable geometries may not be possible, especially in microcolumn techniques. For example, in capillary electrophoresis, on-capillary detection is the norm to reduce band-broadening, and this means introducing the light into a cylindrical tube, one of the worst geometries for reducing scattered light. If considerable scattering of the excitation beam does occur, greater emphasis must be placed on the removal of this light by filtering before the photodetector. Because the fluorescent emission is usually very weak, and under optimum conditions the background light levels should essentially be zero, the requirements for the photodetector are for high sensitivity, and low background noise from the photodetector itself. These requirements are usually met by the use of a photomultiplier tube, rather than the photodiode commonly found in absorbance detectors. The output signal of a fluorescence detector is processed in much the same way as the signal from an absorbance detector, with various degrees of filtering being provided to smooth the output signal. Unlike absorbance detectors which give an output in volts which is directly proportional to absorbance units, fluorescence detectors for HPLC usually just give an output voltage which is related only to the light received at the detector, but which is not absolutely related to a physical characteristic of the analyte molecule such as fluorescence efficiency. Thus while comparison of two absorbance detectors may commence by comparing their specified noise levels in absorbance units, two fluorescence detectors need to be compared using a standard HPLC method to reveal their relative capabilities.

6.4.3 Fluorescence derivatisation

Although many compounds are not themselves strongly fluorescent, it is often possible to react the analyte with a suitable derivatisation reagent to give a fluorescent product. Derivatisation may take place either pre- or post-column. Pre-column derivatisation is quite commonly used; a suitable derivatisation reaction is chosen for the analyte of interest, and the separation is developed using the derivatised analyte. As well as being fluorescent, the reaction product must be stable to allow enough time for the subsequent chromatographic analysis to be performed. Analysis is much easier if only one product results from the derivatisation; multiple tagging of the analyte will probably lead to multiple peaks. Post-column derivatisation allows the use of an already developed separation, with the derivatisation reagent being added to the column effluent. Requirements for a post-column reaction are speed (seconds for complete reaction) and

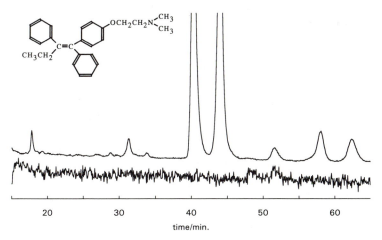

Figure 6.5 Chromatogram of tamoxifen and its major metabolites using fluorescence detection with the post-column photochemical reactor switched on (above) and off (below, with attenuation reduced by 10 times). Insert: structure of tamoxifen. Peaks are 4-hydroxytamoxifen (32 min), tamoxifen (40 min), desmethyltamoxifen (44 min), didesmethyltamoxifen (58 min), tamoxifenol (62 min), others unknown.

lack of fluorescence in the unreacted reagent. Multiple labelling is not a problem as it is with pre-column derivatisation, since no further separation will take place before the detector, and stability is also less of a problem, since the fluorescent product will pass through the detector very shortly after reaction. One limitation of post-column labelling is that the reaction must take place in the presence of the mobile phase, and this will not always be optimum. For either pre- or post-column derivatisation, it is most important to ensure complete reaction of the analytes to obtain quantitative results. A wide range of possible derivatisations have been published, and many of these are reviewed in the reference on derivatisation given in the bibliography.

An interesting alternative to chemical derivatisation is to use a post-column photochemical reactor; this can, for a limited number of compounds, produce highly fluorescent products from an otherwise weakly fluorescent species. The reactor consists of a length of clear HPLC tubing in knitted coil (to minimise band-broadening) wrapped around a UV lamp. The column eluent flows through the reaction coil. Flow rate and length of tubing control the time of exposure to the light. An example of the use of such a system is the analysis of tamoxifen and its metabolites in human plasma. Figure 6.5 shows two chromatograms, one with the reactor turned off, the other with it switched on. The increase in fluorescence of the peaks of tamoxifen and its metabolites is striking.

6.4.4   Operation of a fluorescence detector

This is a little more complicated than using an absorbance detector, since both the excitation and emission properties of the molecule of interest must be considered. Both may change markedly with changes in the solvent composition, so that fluorescence spectra in the literature in different solvents can be taken as an approximate guide only. If a separation method has already been developed for the analytes of interest, the excitation and emission spectra may be determined in the HPLC mobile phase. This information will immediately allow an appropriate choice of excitation and emission wavelengths and bandwidths. Obviously to limit the amount of extraneous light at the detector the excitation and emission bands should not overlap; however, best sensitivity will probably be obtained with the maximum bandwidths which meet this criterion. If fluorescence is disappointingly low it may be worth trying to develop a different chromatographic separation, since fluorescence efficiency can be highly dependent on the mobile phase used. Problems can also occur due to trace levels of fluorescent impurities in the mobile phase leading to high background light levels, and from dissolved oxygen in the mobile phase which can cause fluorescence quenching, necessitating good degassing.

6.5   The refractive index detector

6.5.1   Principles of operation and design of refractive index detectors

The velocity of an electromagnetic wave will vary as it passes from one medium to another; the ratio of the speed in vacuum to that in a given medium is known as the refractive index (RI) of the medium. The RI is a universal property of materials which transmit light, and RI measurements can be used to detect any compound. The HPLC RI detector measures the change in RI of the mobile phase due to the presence of dissolved analyte. When a light wave passes from one medium to another with different refractive indices, both reflection and refraction of the beam occur. In Figure 6.6a, the incident light ray (which has intensity I_0) is shown interacting with a glass–liquid interface, as may be found in a detector cell. At the interface between the two media the light splits into a reflected part which does not enter the liquid, and a refracted part which carries on into the mobile phase, but with altered direction. The angle of refraction is a function of the ratio of the absolute indices of refraction of the glass and liquid. The deflection is given by Snell's law of refraction, which may be written

$$\sin \theta_i / \sin \theta_t = n \qquad (6.5)$$

where the subscript i indicates the incident beam, t is the transmitted

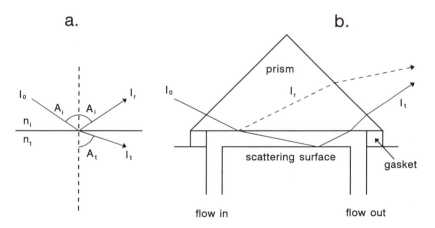

Figure 6.6 (a) Reflection and refraction of a beam of light at an interface between media with refractive indices n_i and n_t. Dotted line is normal to the interface. A_i angle between incident beam and normal, A_t angle between transmitted beam and normal; I_0, I_t and I_r are the light intensities in the incident, transmitted and reflected beams. (b) Cell design for RI detector measuring change in transmitted intensity.

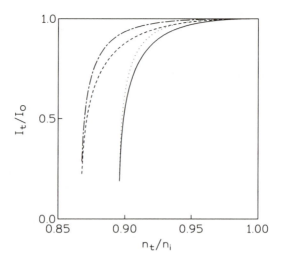

Figure 6.7 Variation of fraction of light transmitted (I_t/I_0) on traversing an RI boundary as a function of the relative RI (n_t/n_i) for angles of incidence of 63.6° (right-hand pair, solid and dotted curves) and for 60.2° (left hand pair, dashed and dash-dot curves). The pairs of curves represent different polarisations of the incident ray, as explained in the text.

beam, and n is the relative refractive index, n_t/n_i. Thus a change in n leads to a change in θ_t, and this can be evaluated by a position-sensitive detector to give a measure of RI.

On passing the interface between the glass and liquid, the light intensity in the incident ray (I_0) is split into two parts, the reflected ray with intensity I_r, and the transmitted ray with intensity I_t, where $I_t + I_r = I_0$. The intensity of the reflected and transmitted rays can be found using Fresnel's laws, and using these equations the curves shown in Figure 6.7 were calculated. Illustrated in Figure 6.7 is the fraction of the incident light intensity which is transmitted (I_t/I_0) as a function of n from $n = 0.9$–1.0, at two different angles of incidence. With $\theta_i = 63.6°$ the two curves to the right are generated, and with $\theta_i = 60.2°$ the pair of curves at the left are calculated. There are two pairs of curves, since the polarisation of the incident light affects the degree of reflection and transmission, and so in each case, there is one curve for light polarised perpendicular to the interface, and another for light polarised parallel to the interface. Transmission reduces to zero when $\sin \theta_i = n$ (the critical angle), while at $n = 1$ the light sees no interface and is 100% transmitted. Clearly the sensitivity (change in transmitted intensity as a function of change in n) is greatest at angles just less than the critical angle, although with lamp sources it is difficult to operate very close to the critical angle because of the limited quality of beam collimation which is possible.

Measurements of either the change in direction or the change in intensity of the refracted ray have been used in HPLC RI detectors. The cell for an instrument which monitors the change in angle of refraction is made up of two triangular section compartments, one of which is the reference cell (filled with mobile phase), and the other the sample cell through which the column effluent passes. The cells are assembled as one unit to minimise temperature changes (and thus RI differences) between the reference and sample sides. Light passes through the sample and reference cells onto a mirror, and is then reflected back through the cells and onto a position-sensitive photodetector. This may take the form of a pair of photodiodes mounted side-by-side, both of which are partly illuminated by the light beam. Any change in the RI of the liquid in the sample cell leads to a change in the refraction of the light passing through this cell, and thus a movement of the light on the photodetectors and a change in the ratio signal measured between them. This signal is then processed and output to the integrator.

A cell for an RI detector which measures the change in transmitted light intensity is shown in Figure 6.6b. As analytes flow through the cell the RI at the interface changes, and thus the transmitted intensity changes. When light strikes the interface at near-normal incidence, almost all of the light is transmitted. However, as the angle of incidence approaches the critical angle, more and more of the beam is reflected, and less passes through the interface (see Figure 6.7). The transmitted light is reflected from the scattering surface, and passes back through the prism to a photodetector. Suitable beam stops are used to reject the reflected beam. The sample and reference cells are located next to each other, separated by a gasket which also defines the distance between the prism and scattering surface. The physical proximity minimises temperature differences between the cells. More than one prism, using glass of different RI may be needed for optimum performance with a wide range of mobile phase RI, although in practice a single prism may given adequate performance for most mobile phases.

RI detectors are very useful instruments because they can be used to quantitate analytes which are otherwise difficult to detect. An example is the analysis of sugars which have poor UV absorbances and thus cannot be detected by UV absorbance or fluorescence measurements without chemical derivatisation. The exception to this universality is when the solute and solvent have identical RIs, whereupon no signal would be observed. The RI of a number of common HPLC solvents is given in Table 6.2.

6.5.2 Operation of a refractive index detector

The universal nature of the RI detector puts this device at a disadvantage in terms of sensitivity, since many factors cause changes in the RI of the column eluent; changes in temperature and pressure are important causes of noise and drift in HPLC RI detectors. Minor pump pulsations which do not cause problems with solute property detectors may cause severe periodic baseline noise with an RI detector. The cause is easy to detect, but the cure may be more difficult. The RI detector should always be present as the last detector if more than one detector is being used. This will minimise back-pressure at the detector and thus minimise the effects of pump pulsation. Another good reason to place the RI detector last is that, with certain cell designs, severe damage to the cell will occur even at relatively low back-pressures due to a second detector connected after the RI instrument. To further reduce pump pulsation noise a pulse dampener might help, or even a change to a pulseless syringe pump if a suitable device is available. Temperature control is another important factor, since the temperature coefficient of RI is of the order of $10^{-3}-10^{-4}$ RI units/°C for many solvents. Thus to be able to measure a change in RI of only 10^{-6} units, the temperature stability between the sample and reference cell compartments must be of the order of 0.01–0.001°C. Heating or cooling of the cell caused by the effluent from the HPLC column can cause great problems; usually RI detectors are built with a lengthy piece of HPLC tubing in thermal contact with the temperature-controlled cell block to allow equilibration of the liquid temperature before it enters the cell. In practice equilibration might not be ideal, and this may lead to drift associated with flow rate changes. A further case of drift in RI detectors is change in the mobile phase composition. This can come about in a number of ways. Mobile phases which are made up of more than one component are best pre-mixed, rather than mixed using a binary pump system; this is not likely to give a sufficiently stable mobile phase composition to obtain a good baseline with an RI detector on a sensitive range. Another problem can be dissolved gas in the mobile phase causing RI changes. Either continuous helium outgassing, or no degassing at all may offer acceptable solutions. Gradient elution operation is practically impossible.

6.6 Electrochemical detectors

6.6.1 Principles of operation and design of electrochemical detectors

There are a number of electrochemical interactions which may be useful as the basis for detection in HPLC; the most commonly used electrochemical detectors are based on amperometric measurements. The principle of operation of an amperometric detector is the oxidation or reduction of analyte in a flow-through electrolysis cell with a constant applied electrical potential, e.g. the oxidation of hydroquinone,

$$\text{(hydroquinone)} \rightleftharpoons \text{(quinone)} + 2H^+ + 2e^-$$

Very low detection limits can be achieved with amperometric detectors (rivalling fluorescence detection), particularly for compounds which are oxidised or reduced at relatively low potentials. In aqueous solvents the limitation on the applied potential is the oxidation of H_2O or the reduction of H^+, at about -1.2 or $+1.2$ V. By limiting the applied potential to that needed to achieve a good response from the target compound, selectivity against compounds with higher oxidation or reduction potentials is achieved. Figure 6.8a represents a block diagram of an amperometric detector.

Noise in electrochemical detectors arises from a number of sources. Background currents may appear as a result of impurities in the mobile phase which can undergo electrolysis. Although a DC background current can be compensated for in the detector offset, it will limit dynamic range, and contribute shot and flicker noise. Pump pulsations may appear as noise spikes on the background current. Non-chemical sources of noise include noise from the detection electronics, grounding problems, poor electrical connections, and pickup of external electromagnetic fields which may require shielding of the detector.

The main differences between electrochemical detectors are the design of the cell used, and the type of electrode material. The electrode material may have favourable characteristics for certain chemical reactions, for example mercury electrodes (whether in liquid form, or with mercury present in a solid amalgam) can be used to detect thiols at a potential of only about $+0.1$ V with the reaction being a complex formation between thiol and mercury. However, for most purposes inert glassy carbon electrodes are used. A variety of cell designs are shown in Figure 6.8b–d. So-called 'low efficiency' designs such as b or c result in reaction of only a small percentage of the analyte, since only a relatively small fraction of the analyte will actually come into contact with the electrode. There are no great advantages for either the thin-layer or wall-jet type of cells; the tubular electrode can also give good results, but it may be rather more complex to manufacture a good polished electrode surface. In the 'high efficiency' design (Figure 6.8d), a porous electrode is used, and almost all of the analyte has a chance to react due to the high surface area of the electrode. Detectors where only a low percentage of the analyte is reacted

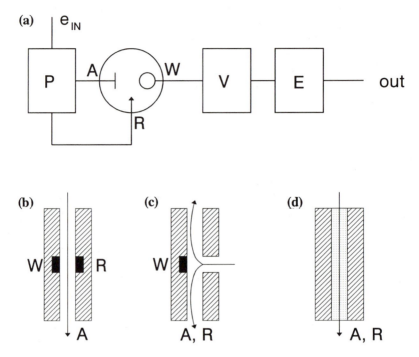

Figure 6.8 (a) Block diagram of an electrochemical detector. The cell has three electrodes. Electrolysis of the analyte occurs at the working electrode, W. An auxiliary electrode, A, supplies current for the reaction, and a reference electrode, R, is used to determine the solution potential. A potentiostat, P, is used to control the voltage at the auxiliary electrode in order to maintain a constant potential difference between the solution (probed by the reference electrode) and the working electrode. The potentiostat is designed so that the reference electrode draws an extremely small current, thus significant voltage drops do not occur within the solution. The current-to-voltage converter, V, takes the current from the electrochemical reaction, and produces an output voltage which is linearly proportional to that current. Added onto these basic building blocks are the usual signal processing modules found in other detectors such as signal filtering, E.

Low efficiency cell designs: (b) thin-layer; (c) wall-jet. (d) High-efficiency cell design with porous working electrode in tube. Arrows indicate liquid flow

are termed 'amperometric', whilst those where almost all the analyte reacts are called 'coulometric'. Although a coulometric design will give rise to a higher reaction current, it will not necessarily give lower limits of detection than an amperometric instrument, since the background current is also proportional to the area of the electrode.

6.6.2   Operation of an electrochemical detector

Compounds such as catechols, phenols, aromatic amines and phenothiazines may be easily oxidised, while quinones and some nitroaromatics are easily reduced. A literature search for the compound of interest, or molecules containing similar functional groups, is likely to be the starting point in looking for a suitable reaction potential. The most suitable conditions may be experimentally determined by observing the detector response over a range of applied potentials. Such measurements can be made with the injector connected directly to the detector (the column need not be in place), so the process can be quite quick and simple.

Slow leaching of electroactive compounds (e.g. iron oxides from stainless steel) from other components in the HPLC system (column, pump, injector) may give rise to high background currents and noise. Therefore, to clean the column before operation it should be pumped with the proposed separation mobile phase for a long time; say, 12 h, or with a strong eluent containing a higher percentage of organic modifier than the proposed separation mobile phase for a couple of hours. To clean the rest of the HPLC system components, a series of flushes using deionised water, then 6 M nitric acid, then glacial acetic acid and finally deionised water for at least 15 min each should be effective; the column should be disconnected, and manufacturer's specifications checked to assess compatibility with strong acids.

The mobile phase itself must have a reasonable conductivity, and usually this means aqueous solutions with a buffer in the 0.01–0.1 M concentration range. The use of non-aqueous solvents with the addition of suitably soluble organic salts is relatively uncommon, but this approach does get around the problem of the limited potential which may be applied with aqueous solutions. For maximum sensitivity, the purity of water, solvents and buffer components is very important to avoid problems due to high background currents. High background currents may also be due to reduction of dissolved oxygen. Recycling of mobile phase may be effective in electrolysing mobile phase contaminants which give rise to background currents, particularly with coulometric systems. A high-efficiency cell may be placed between pump and injector, set at a potential slightly higher than the detection cell to cause electrolysis of mobile phase impurities and reduce the background current at the analytical cell.

6.7   Other detection methods

Many other interactions have been used to probe the eluent from a HPLC column, and only brief mention of a few of these can be included in this volume. The coupling of LC with mass-spectrometry (LC-MS) in particular seems to have great potential and its use will no doubt increase rapidly over the next few years.

6.7.1   Mass spectrometry (MS)

LC-MS is quite widely used, although it is not nearly so popular a technique as GC-MS. The main difficulty in LC-MS is in introducing a sample which is dissolved in a liquid flowstream running at around $1\,ml\,min^{-1}$ into the MS, whilst maintaining a high vacuum within the MS. A number of techniques have been developed for coupling HPLC with MS, and a variety of systems are in commercial production. Others which promise significant improvements in performance are under development.

Direct introduction of liquid into MS systems is possible at low flow rates, say $5–50\,\mu l\,min^{-1}$. Direct liquid introduction (DLI) uses split stream methods to introduce part of the eluent flow from analytical-scale HPLC, but is perhaps best suited to interface the whole flow from a microbore HPLC system to an MS. DLI systems use a probe with a small $(3–5\,\mu m$ diameter) orifice entering directly into the CI source of the mass spectrometer. As liquid passes through the orifice, droplet formation and then evaporation of the mobile phase occurs. A filament in the source causes ionization, and a cold finger may be used to trap excess solvent which may otherwise overtax the vacuum pumping system. The CI spectra obtained with this type of system are most useful for identification purposes. The main disadvantage with DLI systems is poor sensitivity, due to the limited flow rate into the MS. With microbore HPLC less than $10\,ng$ of material injected may give a useful signal, but with split-flow systems up to $1\,\mu g$ may be necessary.

Another popular interface is the thermospray device. In this design, the LC effluent flows into a vacuum region via a heated tube with a sapphire tip through which a small hole has been bored. The temperature and flow rate are adjusted so that the mobile phase is emitted from the sapphire tip as a stable spray of vaporising droplets. To achieve ionisation of the solutes a mobile phase additive such as ammonium acetate or triethylamine is used. The sample ions are accelerated by an electric potential from the thermospray into the MS via a small orifice. The thermospray volume is evacuated by a high flow rate vacuum pump to keep a moderate vacuum in this region, and the small pinhole between this region and the MS results in only a limited gas leak into the system. The gentle form of ionization used in thermospray does not give rise to fragmentation, and so diagnostic information on the identity of a chromatographic peak is limited to the molecular ion, unless MS-MS is utilised. Perhaps the greatest advantage of the thermospray system is that it is compatible with reversed-phase analytical-scale HPLC, with a very high water content in the mobile phase at flow rates of up to $2\,ml\,min^{-1}$. In thermospray, the column eluent is introduced into a low pressure region. Atmospheric pressure nebulization systems have also been developed. Typically a heated nebulizer is used with the column eluent flowing through an inner capillary tube, with an external concentric flow of nebulizer gas. Ionization occurs due to an electrical discharge, and ions are accelerated into the MS via a small hole.

6.7.2   Radioactivity detectors

Radioactivity detectors find quite widespread use in the pharmaceutical industry because of the excellent selectivity they provide when performing pharmacokinetic and metabolism studies with radiolabelled drugs. Decay of a radioactive nucleus leads to excitation of a scintillator material, which then loses its excess energy by photon emission. Photons are counted using a photomultiplier tube, and the number of counts per second is proportional to the level of radioactivity (and thus radiolabelled analyte) inside the measurement volume. The scintillator may be a liquid, which is mixed with the mobile phase post-column, or a solid over which the column effluent flows. The liquid approach has the advantage of allowing the most intimate contact between the analyte and scintillator molecules, and as a result a large proportion of decays gives rise to photon emission. The counting efficiency may be 100 times worse for solid scintillators; however, these systems may offer smaller effective volumes and be somewhat more convenient to use. The counting efficiency and the measured signal are dependent on the type of radioactive nucleus used. For example, ^{32}P has a high energy of β-decay (1.7 MeV) and a short half-life (14.3 days), while for ^{14}C the energy of decay is only 0.16 MeV and the half-life is 5715 years.

6.7.3   Optical activity detectors

Today many new pharmaceuticals and agrochemicals have one or more chiral centres, providing impetus for the recent development of many chiral stationary phases for enantioselective HPLC. Optical activity detectors are capable of specifically detecting chiral compounds, taking advantage of their unique interactions with polarised light. Optical activity detectors are of two types: circular dichroism detectors, which measure differences in absorption of left and right circularly polarised light, and optical rotation detectors, which measure the rotation of the angle of plane polarised light. Circular dichroism is only measured in the absorbance bands of a chiral molecule, and frequently scanning measurements are made to provide a circular dichroism spectrum. This can provide a great deal of structural information, but the process can be rather slow, often requiring stopped-flow operation of the HPLC system. Optical rotation detectors have a universal response for optically active compounds, and so can operate at any wavelength. Optical rotations tend to be larger at lower wavelengths, but the most sensitive optical rotation detectors use laser light sources, operating in the visible (using gas lasers) or even near infrared regions (using laser diodes). In combination with chiral columns, optical activity detectors may be useful in identifying chiral compounds, determining elution orders, and measuring partial enantioselectivity. An optical activity detector may also be used in combination with another detector and an achiral column to determine enantiomeric purity. This is done by comparing the response of the achiral and chiral detectors to the sample, and to a standard of known enantiomeric purity (e.g. a single enantiomer).

6.8 Integrators

The detector gives an electronic output signal related to the composition of the HPLC eluent. It is the job of the last element in the chain of HPLC instrumentation to display and allow the quantitation of the peaks in chromatograms such as Figure 6.1. Until the 1970s (and until much later in many less affluent labs) the strip chart recorder was the main output device, with peak area being determined manually by the analyst; this would be done either by some form of triangulation, or by cutting the peaks of interest from the chart and weighing the resulting paper clippings to obtain a measure of the area. Whatever the method, the analyst's subjective judgement had to be used to decide the proper baseline position; easy with well-resolved peaks and a good signal-to-noise ratio, not so simple with partially resolved peaks of poor signal-to-noise.

Today these functions are usually performed by an electronic integrator which can automatically make determinations of peak areas. This may be a stand-alone device, a computer-based integrator, or a hybrid system where individual fully featured integrators are also networked to a central computer system for long-term data storage. The gain in convenience is enormous; in many cases, the apparent gain in accuracy with peak areas being reported to five or six significant figures is quite illusory. This is because the integrator has to make the same judgements of what constitutes a peak as the analyst with a strip chart, and unless the chromatography is very good, the integrator's best guess will be no better than the analyst's, and frequently worse. So although the integrator can be welcomed as a most significant element in a HPLC system, the results it reports should always be subjected to critical appraisal by the analyst.

The analogue signal from the detector is first digitised by the integrator, typically from 50 to 1000 times each second (the sampling frequency). To have good accuracy in peak area determination, about 100 measurements per peak are desirable. With modern integrators the sampling frequency is adequate even for the determination of capillary GC or capillary electrophoresis peaks with sub-second half-widths. When being used with slower separations the data from groups of samplings are averaged or bunched together to give a similar overall number of digitisations per peak. Bunching is determined by a user-controlled parameter often called 'peak width', and results in a considerable saving in computing time and memory requirements, as well as providing smoothing of short-term noise.

The digitised data are analysed to recognise the presence of peaks, their start and finish, and to calculate peak height and area. This is done by calculation of the first and second derivatives of the signal for small, sequential, overlapping time intervals throughout the chromatogram. For flat baseline, both first and second derivatives are zero. When a peak occurs, the first derivative is positive from the start of the peak to its highest point, then negative until the end. A constantly sloping baseline is distinguished from a peak, since although the first derivative has a non-zero value, the second derivative is zero. The peak width parameter determines the time interval over which a slope determinations are made. Short intervals facilitate the detection of narrow peaks, larger ones the detection of slower peaks. In general there is also an automatic function which allows the peak-width parameter to increase during the run, since later eluting peaks will be broader than early eluters, and one peak width parameter may not be suitable for all. A second important parameter in peak detection is a threshold or sensitivity measurement; this is set such that baseline noise is not registered as peaks. A threshold value can be measured by the integrator, which determines the baseline noise level over a period of a minute or so. This should be done with mobile phase flowing in the system since the noise level may be considerably greater than when the liquid in the flow cell is static. Peak threshold is a different parameter to minimum area or height, which does not affect the integration, but suppresses the reporting of small peaks or long-term baseline noise which have been recognised by the integrator.

There are many standard integrator functions which may be included in a timetable of events to occur during the chromatogram. These include timed autozero functions (to compensate for drift), integration inhibits (to stop integration until after the void volume, or during switching transients in coupled-column systems), functions to control special integration characteristics for certain peaks (e.g. whether to use a tangential baseline for the second peak; draw a perpendicular between two peaks, where the second rides on the tail of the first; or to use forward- or back-projection from a well-defined piece of baseline). With baseline and peak start and end points marked on the chromatogram, the integration can be critically judged. Usually it is possible to modify timed events, and then re-integrate a chromatogram if the original integration was not satisfactory. Alternatively, many computer-based systems allow re-integration of individual peaks 'by hand', often the simplest solution where only a small number of chromatograms are being reprocessed, and when signal-to-noise is poor.

Acknowledgements

I would like to thank Karen Fried and Maria Diez-Perez for supplying chromatograms, and Drs. Anne Aubry and Song Li for their helpful suggestions.

Bibliography

Dyson, N. (1990) *Chromatographic Integration Methods*, The Royal Society of Chemistry, Cambridge.

Lingeman, H. and Underberg, W. J. M. (eds.) (1990) *Detection-Oriented Derivatization Techniques*, Dekker, New York.

Parriott, D. (ed.) (1993) *A Practical Guide to HPLC Detection*, Academic Press, San Diego, CA.

Scott, R. P. W. (1977) *Liquid Chromatography Detectors, Journal of Chromatography Library*, Vol. 11, Elsevier, Amsterdam.

Vickery, T. M. (ed.) (1983) *Liquid Chromatography Detectors*, Dekker, New York.

Yeung, E. S. (ed.) (1986) *Detectors for Liquid Chromatography*, Wiley, New York.

7   Method development and quantitation

W.J. LOUGH AND I.W. WAINER

7.1   Method development

7.1.1   Introduction

In carrying out quantitation of an analyte or analytes in a set of samples, it is first necessary to develop a suitable method. The problem must firstly be defined. It is then necessary to collate and examine all the information that is available regarding the problem. Typically this would include:

– details of the physicochemical properties of the analyte;
– the physicochemical properties of the other components in the matrix;
– features (desirable and undesirable) of the available methods.

7.1.2   Mode of LC and column selection

Method development is often described as an 'art' which can be mastered only by practitioners with many years of experience, but it is also the case that in their simplest form the thought processes that constitute method development may be reduced to an easy-to-follow decision tree that even

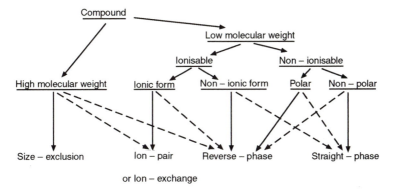

Figure 7.1 Simplified flow diagram for choice of mode of LC. The broken lines indicate that, while there is a single frequently used made for a type of compound/sample, there is usually a variety of other viable alternatives. Less common modes of LC such as ligand-exchange, charge-transfer, inclusion complexation, affinity and hydrophobic interaction are other alternatives. They are first choice methods for some specialised types of sample.

the most inexperienced chromatographer can follow. A simplified decision tree is presented in Figure 7.1 which illustrates that there is more than one option for any given compound/sample type.

Following the guidelines set out in Figure 7.1 would almost certainly lead to an HPLC column and conditions under which there was suitable retention and peak shape for the analyte. At this stage it is necessary to decide upon a sensible column and to persist with it as far as possible, with modifications in the mobile phase being used to improve the method further until eventually it is deemed to be good enough for its intended purpose. In many cases a suitable method will be arrived at without having to reconsider the initial column and/or mode of HPLC choice.

7.1.3   When alternative options must be explored

It is normally not too difficult to find a suitable column which will give desirable retention and peak shape for the analyte. One of the reasons why method development is often difficult is that satisfactory retention and peak shape alone are not enough for a method to be suitable. As discussed in Chapter 1, specificity is very important. In pursuing a separation using one particular mode of HPLC, it may be that, despite all the changes in experimental variables that take place in the method optimisation process, it is not possible to separate the analyte from all interferences present in the sample. In considering the initial stages of method development, it is therefore necessary to bear in mind as many approaches to the problem as possible, in case the first option proves not to be successful.

7.1.3.1 When suitable selectivity is not possible with the initial choice. A case in point is the method development that led to the conditions shown in Figure 7.2 for determining structurally related impurities of a strongly basic drug candidate. The cyano-stationary phase was employed because better peak symmetry was obtained than on an ODS phase, even after looking for improvement by the use of various mobile phase additives. Not all the impurities were resolved from the main peak, however, and most of those that were eluted on the tail of the main peak were not well resolved from one another. Using the ion-exchange phase, all the impurities were resolved from one another and from the main peak, and eluted before the main peak, thus allowing more facile quantitation. This could not have been predicted *ab initio*. Even with prior experience of very similar compounds in similar conditions, it would still be very difficult to predict/anticipate exactly what was going to happen. If time is at a premium, it is prudent to short-circuit the standard route through method development and proceed immediately to the type of system that is known to have a high probability of giving a successful separation for the types of compound under study.

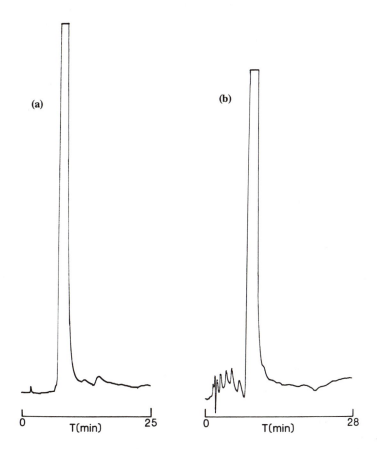

Figure 7.2 BRL 26441A. Optimised separations of main peak from impurity (each at 0.1% by weight) peaks. (a) On Zorbax-CN (DuPont Company, Wilmington, Delaware); co-elution with main peak, or detection problems from broad asymmetric impurity peaks. (b) On Partisil-SCX (Whatman, Clifton, New Jersey).

7.1.3.2 The need for simple, robust methods. Given the difficulty of separating a principal component from structurally similar impurities, it may be advisable to opt right away for conditions that give as much selectivity as possible. For instance, for the analysis of strong bases, good peak symmetry and good resolution of isomeric or closely related species is normally obtainable when using an ODS-silica with a mobile phase containing methanol and an aqueous phosphate buffer at low pH to which is added an anionic ion-pairing agent and a trace of *N,N*-dimethyloctylamine. Also retention and selectivity may be manipulated in a fairly predictable way by changing the ion-pair concentration and, to a lesser extent the pH. Although the success of this method arises from the complexity of the mobile phase, the complexity is also its downfall. This type of method might have some promise as a generic approach to developing methods for strong bases if it were not for the fact that robustness is an important issue for methods which are to be used by many different operators in many different laboratories. While a skilled operator working in one laboratory may be able to reproduce the same types of chromatograms day-in-day-out, it is much better to work with as simple a system as possible to give the method a better chance of transferring well. Therefore a development laboratory producing methods to be used in production quality control might, at the expense of increased method development time, produce methods using a wide range of differing types of HPLC column but with each method using a simple mobile phase, rather than aiming for easily developed methods all using a complex mobile phase but with the same type of column.

Another ironic feature of the increasing emphasis on robustness to ensure ease of method transfer is that the 'quality' of HPLC stationary phases is improving. The quality arises from better batch-to-batch reproducibility and more homogeneous surfaces on which only one type of retention mechanism may operate. The irony lies in the fact that in the past it might have been the 'flaw' (e.g. residual silanols on a reversed-phase material) that was responsible for the last bit of selectivity that was needed to achieve a very difficult separation. Now with 'improved' stationary phases, there may be a need for more complex mobile phases thereby compromising robustness once more.

There is further reason for the type of schematic shown in Figure 7.1 to be of value only as a starting point. The most obvious choice from the schematic may well not be the preferred choice because of conflicting specific local requirements. One such situation has already been mentioned in another context, i.e. there may be a policy in a laboratory to use the same HPLC packing material, as far as possible, for all methods used in the laboratory.

7.1.3.3 Local considerations. There are many other local factors which dictate the path of method development. An increasingly common example is the development of a method which, as well as serving for quantitation, may be directly transferred to HPLC-mass spectrometry (commonly referred to as LC-MS) to allow the identification of minor components in a sample. This is achievable if the desired separation can be obtained using an ammonium acetate buffer instead of the more usual phosphate buffer. On the other hand, if it was intended that the unknown minor components should be isolated in milligram quantities to allow full structural elucidation using techniques such as ^{1}H NMR as well as mass spectrometry, then conditions would be developed especially for the isolation unless the conditions for quantitative work coincidentally also lent themselves to facile scale-up for quantitation. Reversed-phase conditions may be used for preparative work but on the whole it is better to employ straight phase conditions employing 'bare' silica; the packing material is cheaper, the mobile phase is more easily removed, excellent selectivity is often found and back pressures at high flow rate are lower than when using aqueous mobile phases. Peak tailing with polar basic compounds is, of course, a problem, but this can be reduced to acceptable levels by using a relatively volatile mobile phase additive such as triethylamine. In other words, polar compounds can be chromatographed by straight phase HPLC. This is a good illustration of why the schemes such as that shown in Figure 7.1 and the computer-aided learning method development packages based on the same principles should be approached with caution.

7.1.3.4 Avoiding gradients. Another specific instance when it would be necessary to look beyond the immediately obvious choice of mode of HPLC is medium polarity compounds between which there is a wide range of polarity. In using reversed-phase HPLC it might be necessary to tolerate a long chromatographic run time or else adopt gradient mobile phase conditions. As long as the compounds are ionisable this dilemma may be avoided by using an ion-exchange system in which the salient properties of the compounds will be more like one another, thus allowing isocratic elution in a short run time given the obvious caveat that there is still sufficient difference between the compounds to allow separation.

7.1.4 Chromatographic 'expertise'

It is clear that for successful method development, the chromatographer needs to have as many potential solutions to the problem as possible at his/her disposal. There may be some very fine judgments to be made between different methods and therefore there is a need to be aware of the subtle idiosyncrasies of each type of method. Just a few illustrations of the kind of knowledge that might come into this category are: (i) for some older types of cyano-phase there may be large differences in k' in switching from column to column if there is a low percentage of organic modifier in the mobile phase (this presumably arises from poor batch-to-batch reproducibility); (ii) if cyano-phases are exposed for a long time to aqueous buffers at low pH, it is possible that they may be hydrolysed to give some carboxylic acid groups; (iii) conditioning times are shorter in ion-pair HPLC when using a high concentration of a medium length chain ion-pairing agent than a lower concentration of a longer chain ion-pairing agent; (iv) robustness is a problem when using silica under straight phase conditions but if it is used as an ion-exchanger with aqueous eluants, the reproducibility of k' values is surprisingly good; (v) aqueous mobile phases containing a low % organic content should be avoided if the method is to be transferred directly to a coupled LC-mass spectrometry (LC-MS) system which uses one of the more modern (electrospray) LC-MS interfaces.

Given such minutiae and the added volume of detailed information that might be built up with experience relating to peculiarities of the behaviour of each compound type with each type of stationary phase, it is small wonder that it is very difficult to set up a computer-based 'expert system' which will be genuinely useful in method development and in some quarters method development is still considered to be an art form. Even with the best of expert systems there is every chance that in the next problem it is asked to tackle it will be confronted with a complex structure unlike any it has previously encountered. In such situations the 'chemical intuition' of the experienced chromatographer would hopefully provide an advantage. Such deficiencies in computer-based expert systems are being addressed by work on neural networks. These are effectively expert systems which are capable of learning from mistakes and previous experience. They are not yet in common use but in years to come may make decision making in method development much easier. This is especially the case for fields such as pharmaceutical research where new compounds bear a distinct resemblance to those that have gone before them.

7.2   Optimisation

7.2.1   Introduction

Once a mode of HPLC has been selected, and also usually after a column and conditions have been determined under which there is reasonable peak shape and retention for the main peak, the process of optimisation begins. This generally involves manipulation of experimental variables relating to the mobile phase until the desired separation has been achieved.

7.2.2   What is the optimum?

These variables may include the nature and percentage of organic components, pH, ionic strength, nature and concentration of mobile phase additives (e.g. ion-pairing agent, masking agent) and temperature. These may be all optimised simultaneously, as for example in Simplex optimisation one of the earliest optimisation techniques to gain common acceptance. In Figure 7.3 the progression of a modified Simplex optimisation is illustrated for the optimisation of two variables. In this particular example, the separation of two peaks was being studied but it is important

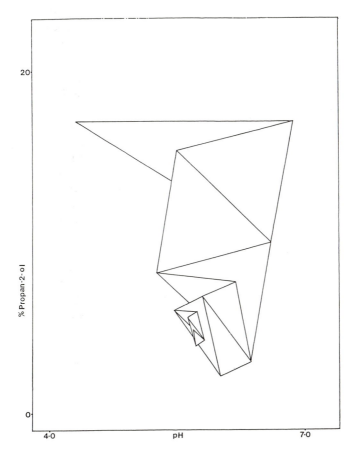

Figure 7.3 Progression of a modified Simplex optimisation. Three initial experiments are required for optimisation using two variables. In Simplex optimisation the next experiment is selected by reflecting the worst vertex through the midpoint of a line joining the other two. In the modified approach, progress towards the optimum may be made in fewer experiments by contracting or extending the nextpoint away from the line depending on how close the worst result was to the other two experiments. (Reprinted from T.A.G. Noctor, A.F. Fell and B. Kaye (1989) Optimisation, in *Chiral Liquid Chromatography*, ed. W.J. Lough, Blackie, pp. 235–243 with permission).

to note that in such cases it is not often simply a case of optimising the resolution between the two peaks. It is necessary to create a 'chromatographic response factor (CRF)' and it is then this number which is optimised. The case illustrated in Figure 7.3 was approached using

$$CRF = (R_s/1.5)^2 - 0.25(T_L - 16) - 0.05(7 - T_F) \qquad (7.1)$$

where T_F is the time to the first peak and T_L is the time to the last peak, with the factors 0.25 and 0.05 being established empirically.

At first this may appear complicated but the last two terms are present simply to put a bias towards solutions in which resolutions take place in 7–16 min and the factors are a measure of the relative importance of staying within the individual time limits. If the first peak elutes in under 7 min it may co-elute with non-retained or weakly retained interferences while if the second peak elutes after 16 min the analysis time will be becoming too long.

A similar CRF may be used for the separation of more than two peaks if it is known that there will always be one critical compound which elutes closest to the main compound of interest. This may be the case for example in a method for determining the purity of a compound when there is one impurity which is very closely structurally related to the compound of interest and therefore much more difficult to resolve from the peak for the main component of the sample than anything else. In other cases, however, it may be equally important to separate all the components from one another. In such instances there must be a high weighting on solutions where the number of peaks observed is the same as the number of components in the sample. Another feature of such examples is that when experimental variables are changed, the elution order of the compounds may change. In some optimisation procedures this is ignored but more frequently the peaks are 'tracked'. This may be done, for

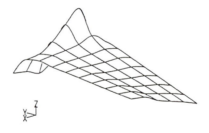

Figure 7.4 A typical response surface map $x = \text{p11}$, $y = $ percentage of propan-2-ol, $z = $ chromatographic response factor; orthogonal views. (Reprinted from T.A.G. Noctor, A.F. Fell and B. Kaye (1989) Optimisation, in *Chiral Liquid Chromatography*, ed. W.J. Lough, Blackie, pp. 235–243 with permission).

example, by using a UV diode-array detector, which will simultaneously detect at all wavelengths over a pre-defined range during the course of a chromatogram. A peak may then be identified by using the UV spectrum of the compound responsible for the peak or an absorbance ratio for two wavelengths as its 'signature'.

Returning to the two component system used as an example in Figure 7.3, the optimisation proceeds across the total response surface (e.g. Figure 7.4) until the maximum response is found. There may be a sharp slope of the response surface close to the maximum response in which case marked improvements in the separation would be seen as the optimum was approached. However, ideally the optimum would be approached more gradually with a shallow slope around the maximum. In this way a method based on the optimal conditions would be robust, i.e. not sensitive to minor changes in experimental variables.

7.2.3 Local versus global maxima and consideration of number of experiments

Such an optimisation using two variables may be conveniently represented diagrammatically and, with the aid of a computer, the procedure can be applied to the simultaneous optimisation of multiple variables. For example if there are six variables, then seven initial experiments are required before the sequential optimisation may begin. Although the end results of such multivariate optimisations may be impressive, they have their limitations. Centring on a local optimum is a potential pitfall in all optimisations but is more so when adopting a sequential approach, especially with a large number of variables. A more important limitation, however, is that more experiments than are actually necessary are likely to be carried out. Several of the variables may have no or only a very limited effect with respect to reaching the desired separation. It is therefore more sensible to carry out some kind of initial factorial design experiment to determine which variables are the most important and then proceed with an optimisation based on these fewer, more important variables. Although this approach makes the fairly presumptuous assumption that the effect of each variable is independent of the others, it is generally accepted as being a more sensible way of approaching things.

7.2.4 Computer-aided optimisation

For optimising two parameters at a time, one of the simplest, and easy to use computer software packages is DryLab, produced and marketed by LC Resources. Data from just two runs is used to predict both isocratic and gradient conditions. Usually the two runs are gradients in which the % organic component in the mobile phase is varied with time. The peak information from the two chromatograms generated is then entered into the DryLab computer program. The program then simulates both isocratic and gradient runs until the optimum separation is achieved. It is then of course necessary to run the optimum set of conditions on the HPLC system to verify the accuracy of the prediction! One obvious disadvantage of this system is that if it is not possible to obtain a separation by manipulating the two variables then DryLab will not find one. Bearing this in mind, for very difficult separations DryLab may be used initially more as an aid to method development to ensure that conditions are found under which the compounds being analysed have suitable retention times before exploring further using subtle variations of other variables.

Most benefit can be had from computer-aided optimisation if it can be carried out totally unattended. This requires highly automated instrumentation and therefore it is no coincidence that the software packages in common use have been developed by instrument companies for use with their own instrumentation. The programs not only make predictions for the next set of conditions, but also set up and run these conditions on the HPLC system. Such computer packages include ICOS (Interactive Com-

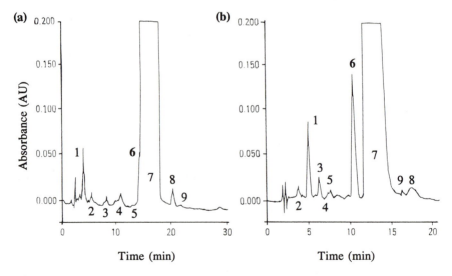

Figure 7.5 Optimisation of a separation of degradation products produced in a stability trial. Originally the separation of the active agent and a number of its impurities and degradation products was achieved using a complex ternary gradient over 30 min. Using Diamond, an isocratic method was developed taking only 20 min to run. This reduced the analysis time and removed the need for re-equilibration between runs. The method also had better resolution, as one of the major degradation products was completely resolved from the major component.

puterised Optimisation Software, from Hewlett Packard), PESOS (Perkin Elmer Solvent Optimisation System, from Perkin Elmer), DIAMOND (from Unicam), MILLENIUM (from Waters) and EMERGO (from Chrompack). There are many differences between these programs but they are similar in that to varying degrees, unlike Simplex optimisation, they explore the whole mobile phase composition space as well as progress towards the optimum separation.

The chromatograms shown in Figure 7.5 illustrate the degree of improvement that can be achieved by using the DIAMOND software. This is one of the most complete of the optimisation systems and uses most of the attractive features found in other systems. Firstly it rapidly establishes a suitable isoelutropic plane in the tetrahedral factor space generated by three organic modifiers and an aqueous constituent. At all solvent compositions on this plane, the last peak in the chromatogram will have the same retention time. A small number (i.e. 10) of experiments are then conducted on this isoelutropic plane. The use of a mathematical technique called piece-wise quadratic modelling then allows the entire isoelutropic plane to be mapped. From this the user can select the type of separation they require, e.g. the most even separation between peaks or the maximum separation between the worst resolved pair of peaks. The speed with which the optimum is reached benefits from the use of a diode-array detector to track the peaks (i.e. identify the peaks when changes in retention order take place). Also the accuracy with which retention times under certain conditions are predicted is enhanced by the use of peak deconvolution of overlapping peaks found in any of the ten experiments so that the k' or retention time can be measured with greater certainty.

7.3 Relative assays

For accurate results the purity of a compound or the % content of a compound in a product is determined by a relative assay in which a sample of unknown purity or content is compared against a pure reference standard. The only exception to this is when a compound has just been discovered and no suitable reference standard is yet available.

7.3.1 External standard method

In the external standard method the detector response for pure compound is found by determining the peak area per unit concentration of pure compound by obtaining chromatograms from injections of solutions of the reference standard at known concentrations and measuring peak areas. This detector response may then be used to calculate how much pure compound is in a solution of a sample by considering the peak area obtained from the injection of a sample solution of known concentration.

This method involves the usual sources of error such as those made in measuring weights and volumes. Two key potential sources of error are those occurring through poor repeatability of injection volumes and through inconsistent loss of sample in any sample preparation step. Both of these may be compensated for by the use of an internal standard method.

7.3.2   Internal standard method

The internal standard method is similar to the external standard method in that solutions of the reference standard are compared with solutions of the sample. The key difference is that prior to any sample pre-treatment all solutions are 'spiked' with the same amount of a compound called the internal standard. For this method to work well, it is important to choose a suitable internal standard. Ideally an internal standard should:

- be recovered to a similar extent to the analyte in any sample pre-treatment procedure;
- be reliably resolved from the analyte peak and any other peaks that might reasonably be expected to arise from other components in the sample;
- have a detector response factor similar to the analyte;
- be present in a similar concentration to the anticipated concentration of the analyte in sample solutions;
- by its presence not give rise to an unacceptable increase in the chromatographic run time.

From these considerations it is clear that the internal standard should have very similar physicochemical properties to those of the analyte, yet not so similar that it cannot be resolved from the analyte in the chromatographic system. In many situations it might be difficult to find a compound to fit all these criteria. An exception to this is in the analysis of drug candidates in development. In such cases there would almost definitely be a range of compounds with very similar structures readily available as a result of the drug discovery programme. An illustration of how an internal standard might be chosen is shown in Figure 7.6.

Because the internal standard method eliminates some of the errors found in the external standard method it does not automatically follow that the internal standard method should always be used. The precision of many LC external standard methods is very good (e.g. <0.4% RSD for a purity determination) given that (i) the repeatability of injection volumes in modern injectors is much better than it used to be, especially if the injection is automated and (b) there are many methods for which sample

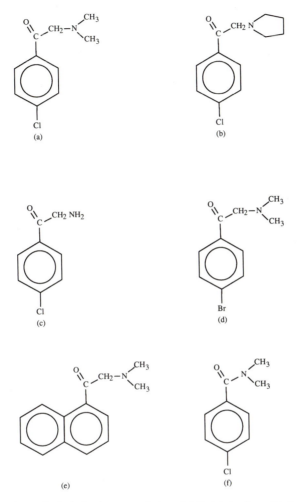

Figure 7.6 Internal standard choice in reversed-phased LC. (a) Analyte. (b) Suitable internal standard if readily available; similar physicochemical properties to analyte but probably enough difference in hydrophobicity to allow separation. (c) Unsuitable, different pK_a from analyte, therefore large difference in k' at some pH values. (d) Unsuitable, not the best match for λ_{max} and likely significantly larger k' values. (e) Highly unsuitable, very different molar absorbance. (f) Highly unsuitable, different behaviour relative to analyte with changes in pH (i.e. neutral of basic analyte).

preparation is simple, involving no loss of the analyte. For these cases there is no advantage in using an internal standard method especially since it involves an extra step and extra measurements. Moreover there are cases where the use of an internal standard would actually lead to poorer precision. This would be the case if the LC peaks were tailed and the major source of error was in the estimation of peak area. The external standard method involves one peak area measurement per chromatogram while the internal standard method involves two. Hence there is greater error in the internal standard method ($\sigma^2_{total} = \sigma^2_1 + \sigma^2_2 + \sigma^2_3 + \cdots$).

7.4 Method validation

7.4.1 Introduction

Having developed a separation of analyte(s) from all possible other components that are likely to arise in the samples being analysed and decided how the quantitation will be carried out, the proposed analytical method must be validated before it can be used. In other words, it must be demonstrated that it is suitable for its intended purpose by carrying out a series of tests. The reason for carrying out these tests is often given as being to satisfy the demands of regulatory bodies. However while these demands were the catalysts for the trend in recent years for analytical methods to be thoroughly validated, they have merely served to raise awareness. It is a simple fact that the results of any analytical method are worthless if the method is not suitable for its intended purpose.

The validation process begins in method development in that the documentation reporting the validation data must include a record of the method development process, giving details of the conditions explored and the rationale of the progression of the process. The validation proper consists of a series of tests for which there are acceptance criteria which vary depending on the type of assay being carried out. Literal definitions of the parameters for which tests must be carried out are discussed elsewhere in this book but are repeated here as a reminder.

7.4.2 Specificity

Since achieving specificity is often the most difficult aspect of developing an assay, establishing specificity is inextricably involved in the method development process. Therefore when it comes to validation, it should be simply a case of demonstrating specificity. It may be the case that when more thorough checks are made as in the validation specificity tests, the method is not as specific as was first thought.

An assay is specific if the 'analytical response' (i.e. that which is measured) arises from the analyte of interest and cannot arise from any other compound likely to be present in the sample. Therefore for an HPLC method, unless a selective detector is being used, it is necessary to demonstrate the peaks arising from all other compounds likely to be present in the sample are at least baseline-resolved from the main peak.

For all such likely other compounds for which specimen samples were available, k' would be determined under the conditions of the HPLC method to demonstrate that it was different from k' of the compound being determined. Specimen samples of likely interferences are not commonly available. Then it is necessary to obtain a chromatogram of a typical sample or better still a chromatogram of a sample which is thought to be enriched in likely interferences. For example, for a method to determine purity, a chromatogram of a highly impure sample would be obtained. For acceptable specificity there should be no peaks that are only partially resolved from the main peak. Also the main peak should be 'homogeneous'; i.e. there should be no peaks underlying the main peak.

Establishing peak homogeneity is no easy matter. It may involve studying the sample using a variety of different HPLC conditions and/or using techniques such as thin-layer chromatography (TLC) and proton nuclear magnetic resonance spectroscopy (^{1}H NMR) to search for additional impurities. There is some assurance of specificity if all the impurities found by the alternative methods are accounted for in the chromatogram obtained using the conditions of the method that is being validated. In other words, if they can be observed in the chromatogram then they cannot be hidden under the main peak. UV diode-array detection may also be useful in checking peak homogeneity. The UV spectra taken at various points on the peak should be a good match (there are various ways of testing the degree of match) for the UV spectrum of the pure compound. If the UV spectrum of a pure reference standard is not available, then at least the UV spectrum across the peak should not change. If there is partial resolution under the main peak then the UV spectrum or absorbance ratio of two or more wavelengths would change from the front edge to the apex to the tailing edge of the peak. This is quite a discriminating test if there is a co-eluting compound which has a significantly different UV spectrum from the compound being determined. However, if the UV spectrum of the interference is similar to that of the compound being determined, as is frequently the case with co-eluting structurally related interferences, then quite large amounts (up to 10–20% by weight) of interference may be 'missed'. In time, with the advent of low cost LC-MS (HPLC coupled directly on-line with a mass spectromonitor), peak homogeneity testing will be much less problematic. Mass spectra of even closely related compounds are distinctly different and accordingly it should be possible using LC-MS to detect very low levels ($<0.1\%$ by weight) of interferences under the main peak.

The acceptance criteria for specificity, except in methods which use selective detection, is almost always 'reliable baseline resolution' of the main peak from all peaks arising from likely interfering components of the sample. The only difference between different types of method for different types of sample is in the nature of the potential interferences.

7.4.3   Robustness

A robust or rugged method is one which is not adversely affected by minor changes in experimental variables of the order that might reasonably be expected to take place during the course of the operation of the method. This is therefore tested by carrying out the chromatography on columns of the same type but with different histories and/or using different batches of mobile phase, perhaps even intentionally making some minor changes in composition. In principle when checking for robustness in this way, the acceptance criteria should be that over these different runs all the other validation tests are still passed. However in practice it usually suffices to establish that resolution from a critical close-running peak is maintained.

7.4.4   Linearity

A method is linear if there is a linear relationship between the 'analytical response' and concentration of analyte in the sample solution over a specified range of concentrations of the analyte. The plot of 'analytical response' versus analyte concentration should have negligible intercept.

With respect to an HPLC method, the 'analytical response' is almost always based on peak area. Peak heights may be used when working close to the limit of detection but are otherwise not used since if there is any peak asymmetry present a plot of peak height versus concentration will deviate from a straight line.

The range over which linearity is tested depends on the type of method. Linearity is generally assessed by calculating a linear regression coefficient (r) and, although on occasion it might be acceptable to allow a lower figure, it would usually be expected that $r > 0.999$. Whether or not an intercept is negligible depends on its size relative to the 'analytical response' or concentration of typical samples. For instance, if the intercept was on the y-axis, it might be expected that the intercept expressed as a percentage of the peak area for a typical sample was $<0.5\%$.

7.4.5   Precision

The precision of a method is the degree of closeness of the results and is usually reported as a % relative standard deviation. It is often subdivided into repeatability and reproducibility. Repeatability is assessed by making replicate injections of the same solution and gives a measure of the error in measuring peak area and, if an internal standard is not being used, in the injection volume. For the %RSD figure obtained to be ideal in terms of statistical significance, it would be necessary to make quite a large number (in the order of 20) of injections. However, as a compromise to save time, it is quite usual to make 6–10 repeat injections. Similarly there are pragmatic compromises that may be made in assessing reproducibility. What is ideally required is to carry out replicates of the complete assay and find the %RSD on the result. As well as the errors that are assessed in measuring repeatability, this gives a measure of the error in preparing sample solutions. A similar measure may be had by calculating the %RSD on the detector response for the analyte (or detector response relative to an internal standard) based on 6–10 individual injections of different sample solutions. Therefore it is often deemed to be sufficient to carry out this shorter experiment.

7.4.6 Accuracy

An assay is accurate if the mean result from the assay is the same as the true value (i.e. there is no bias (systematic error)). Assessment of accuracy is notoriously difficult since often the true value is not known but, generally, good accuracy is a consequence of the other validation parameters being within acceptable limits.

To test accuracy for an HPLC method it is necessary to carry out the method on a sample of known composition. This presents a problem since the composition cannot be known unless there is an alternative method which has already been validated. This is particularly a difficulty for methods for determining purity since this alternative method would be needed to assign the purity of the reference standard. In the absence of a reference standard of assigned purity by an alternative validated method, the purity of the reference standard must be assigned on a 100% detected impurities basis. At best the purity assigned in this manner will be a good approximation and there will be a loss in accuracy in using the new HPLC method that is to be validated. Since this loss of accuracy arising from error in assigning the reference standard purity cannot be distinguished from any inaccuracy inherent in the HPLC method, a proper assessment of accuracy cannot be made.

As already indicated, this issue of obtaining an accurately assigned purity for a reference standard is important in purity determinations. Incorrect assignment of the reference standard purity leading to the sample being assigned, say, as 99.8% pure instead of 99.2% pure may be critical, especially if the safety of the compound for human consumption is being considered. However, with respect to methods for determining the content of a compound in a product, biological fluid or environmental sample, the relative systematic error (bias) introduced by incorrect assignment of the purity of the reference standard will be fairly small compared to any likely relative systematic error in the HPLC method so long as the content of the compound in the sample is lower than about 20%. It is therefore satisfactory to proceed to check the accuracy of the method by using it to determine the content of a sample of known content. This sample of known content (reference sample) is prepared by adding ('spiking') a known amount of compound to a known amount of the sample matrix which contains no compound (i.e. 'blank' matrix). Obviously when dealing with solid samples some consideration needs to be given to the mixing of the reference sample and to the degree of intimacy of the compound with the matrix (e.g. on standing, adsorption of the compound onto matrix components may occur and this may affect the release of the compound into free solution during sample preparation for the HPLC method).

7.4.7   Limit of detection

The limit of detection (l.o.d.) is the amount of analyte which can be reliably detected under the stated experimental conditions. Although a statistical approach may be adopted in defining exactly what is 'reliable' and what is not, it is common for individual analytical groups/departments to have a pre-determined policy (drawn up, of course, taking due

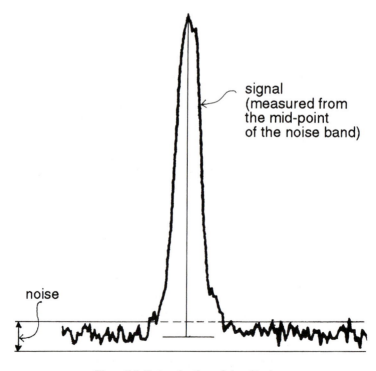

Figure 7.7 Determination of signal/noise.

account of statistical significance) on what they consider to be the l.o.d. This will frequently be the concentration or weight of compound that will give rise to a signal-to-noise ratio of 2 (or 3). The signal is the peak height and the noise is the amplitude of short term noise in the baseline as shown in Figure 7.7.

7.4.8   Limit of quantitation

The limit of quantitation (l.o.q.) is the amount of analyte which can be reliably quantified under the stated experimental conditions. It is equivalent to a higher concentration or weight than the l.o.d. and accordingly is usually taken as some predetermined multiple of the l.o.d. This might typically be in the range of 3–5 and almost always be equivalent to a signal-to-noise ratio of at least 10. Alternatively, given that precision decreases with amount of analyte injected and precision is critical with respect to quantitation, l.o.q. may be defined in terms of the lowest amount of analyte that will give a method %RSD that is within predetermined limits. What these limits are will depend on the type of method. As an illustration, for the determination of a drug in a biological fluid the l.o.q. may even be taken as an analyte concentration giving a %RSD as high as 20.

7.4.9   Stability in solution

It is necessary to study stability in solution in the solvent used to prepare sample solutions for injection in order to establish that the sample solution composition, especially the analyte concentration, does not change in the time elapsed between the preparation of the solution and its analysis by HPLC. This is a problem for only a few types of compound (e.g. penicillins in aqueous solution) when the sample solution is analysed immediately after the preparation of the sample solution to be injected. The determination of stability in solution is more of an issue when sample solutions are prepared and then analysed during the course of a long autosampler run. While the acceptance criteria for stability in solution may be expressed in rather bland terms by making a statement such as, e.g. 'the analyte was sufficiently stable in solution in the solvent used for preparing sample solutions for reliable analysis to be carried out', in practice it has to be shown that within the limits of experimental error, the result of the sample solution analysis by the HPLC method is the same for injections at the time for which stability is being validated as for injections immediately subsequent to the sample solution preparation. While this may be done by a subjective assessment of results with confidence limits, strictly speaking a statistical method known as the Student's t-test should be used.

7.4.10 System suitability

The design of a system suitability test should not really be deemed to be a part of method validation but it is nonetheless a very important part of ensuring that the method fulfils its intended purpose. Once an HPLC method has been validated by virtue of the results for the tests described above fitting the acceptance criteria, it is still necessary each time the method is used to check that it has been set up according to the documented description of the method and, inherent in this, that valid results are obtained. While the analyst would like a quick check to ensure that the method has been set up correctly, there is no escaping the fact that in theory what would really be needed to ensure that everything was in order would be to carry out the whole battery of validation tests before, during and after the use of the method. Clearly this is impractical since only a small fraction of the run time would be spent actually running samples. Therefore in practice short system suitability tests are designed but every

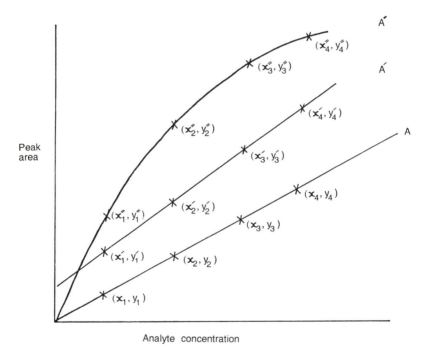

Figure 7.8 Rationale for recommended system suitability test. A, linear, negligible intercept, response $= y_1/x_1 = y_2/x_2 = y_3/x_3 = y_4/x_4$ (therefore RSD on response measured at different analyte concentrations should not be significantly higher than the precision of the assay based on the concentration at mid-point of linearity range provided the different concentrations are not too far from the mid-point); A', linear, non-zero intercept, $y'_1/x'_1 \neq y'_2/x'_2 \neq y'_3/x'_3 \neq y'_4/x'_4$; A", non-linear, zero intercept, $y''_1/x''_1 \neq y''_2/x''_2 \neq y''_3/x''_3 \neq y''_4/x''_4$.

effort is made to extract as much information as possible from these short tests. A system suitability test involves the injection of a sample solution containing analyte spiked with an important close running impurity. Retention, resolution, efficiency and peak symmetry are measured and compared to the expected values. This is convenient since many modern HPLC systems measure these parameters automatically. The convenience may in fact be the overriding factor since it could be claimed that at least two of the parameters are of little value; poor efficiency and bad peak tailing do not make for pretty chromatography but could perhaps be tolerated so long as resolution is maintained. A more valuable system suitability test is one in which three (or four) analyte solutions, spiked with impurity, are injected once each, and resolution and detector response (peak area/analyte concentration) for each solution are measured. The former gives a quick check on specificity while the latter gives a quick check on both precision and linearity by looking at the %RSD on the detector responses. For a straight line calibration plot with negligible intercept, all points on the line should have the same detector response which is equal to the gradient of the straight line (Figure 7.8).

7.4.11 Method transfer protocols

Difficulties in using HPLC methods frequently arise when a method is developed in one laboratory and is then transferred for use in another laboratory. To overcome this problem, there is a current trend towards developing method transfer protocols to facilitate a smoother transition. These protocols not only involve additional testing but also may even involve visits (for observation and supervised introduction to carrying out the method) of staff between sites. This seems over elaborate and it remains to be seen whether this trend will continue given that it should not be necessary if a method has been rigorously documented and validated.

7.5 Types of quantitative method

7.5.1 Assignment of purity

As discussed earlier, the assignment of purity is ideally done by a relative method, comparing analyte peaks in chromatograms of sample solutions with analyte peaks in chromatograms of a reference standard of assigned purity. The difficulty that arises when no suitable reference standard exists has already been alluded to. Purity then has to be assigned on a 100% detected impurities basis. The nature of the difficulties that arise in assigning purity in this way is discussed in the next section.

When carrying out the relative LC method there is no need to carry out a calibration plot with standard solutions since the method will already have been demonstrated to be linear during validation and perhaps also by the system suitability test. It is preferable to prepare all the standard and sample solutions at approximately the same concentration (although clearly the exact concentration of compound in the sample solution is not known since this is what is being determined!). In this way error is reduced by comparing like with like. A concentration in the middle of the range is chosen, also with a view to reducing error. Following injection of the sample and standard solutions and measurement of the peak areas in the chromatograms, the % purity of the sample may be determined from

$$\% \text{ purity} = \frac{A_1}{c_1 \times r} \tag{7.2}$$

where A_1 is the area of the analyte peak in the sample chromatogram, c_1 is the concentration of sample and r is the average response from relevant reference injections; for each reference injection,

$$\text{response} = \frac{A_2}{c_2 \times p} \tag{7.3}$$

where A_1 is the area of analyte peak in the reference chromatogram, c_1 is the concentration of reference standard and p is the % purity of the reference standard.

The use of the phrase 'relevant reference injections' is highly relevant! Ideally under fully equilibrated LC conditions the response would be identical for all injections of reference solutions. In practice there may be a slow long term change (drift) of the response. To account for this each sample solution or sets of sample solutions are 'bracketed' by standard solutions. At the end of the run the analyst must decide whether it is more appropriate for any sample solution to use the average response over the whole run or the average response from standard solutions immediately before and immediately after the sample solution. To be able to quote the sample purity with any degree of statistical confidence, a minimum of 5–6 injections of sample solutions should be carried out.

7.5.2 Determination of impurities

As for HPLC purity methods, the determination of impurities in a sample of analyte is best done by a relative method comparing impurity peaks in sample solution chromatograms with impurity peaks in chromatograms from solutions of reference standards of impurities. Unfortunately, however, it is often the case that such reference standards are not available. Under these circumstances some approximations must be made.

The assumption is often made that the impurities have the same detector response of the main component of the sample. This assumption is implicit in the common practice of expressing % impurity as peak area of the impurity peak as a % of the total area of all peaks detected. This method is adequate for impurities in excess of 0.5% but has shortcomings for impurity levels below this. Trace impurities might only be detectable if a more concentrated sample solution is injected. However, if this is done, the main component may be at a concentration outwith the linear range. If this happens a low area will be recorded for the main peak and consequently the result for the impurity will be too high.

An alternative method which avoids this problem is one in which a 200-fold dilution of the concentrated sample solution is prepared. This solution is then used as a reference solution for all the impurities detected in the concentrated solution. The content of each impurity in the sample is calculated as follows:

$$\% \text{ impurity} = (A_1 \times c_2)/(A_2 \times c_1) \times 100 \tag{7.4}$$

where A_1 is the peak area of the impurity peak in the sample solution chromatogram, A_2 is the peak area of the main peak in the diluted solution chromatogram, c_1 is the concentration of the main component in the sample solution and c_2 is the concentration of the main component in the diluted solution.

Despite the extra error involved in using two solutions instead of one, this method is still much more suitable for trace impurities than the area % method. The ease with which the impurities may be detected in comparison with using a solution for which the main component is in the linear range is illustrated in Figure 7.9. It should be noted, however, that this method is not suited for highly impure samples since the calculation assumes that the concentration of the main component in the diluted solution is effectively the same as the sample concentration.

7.5.3 Content determination

For the determination of the content of a compound in a mixture (product, formulation, biological fluid, soil sample, etc.) a reference standard will be available almost without exception. The main issue with respect to calculations is that a dilution or concentration factor will usually need to be taken into account since sample preparation is often used not only for clean-up but also to ensure that the concentration of the analytical solution injected is above the limit of quantitation. The concentration of the compound in the original sample solution may be calculated using

$$\text{content} = A_2/A_1 \times c_1 \times f \tag{7.5}$$

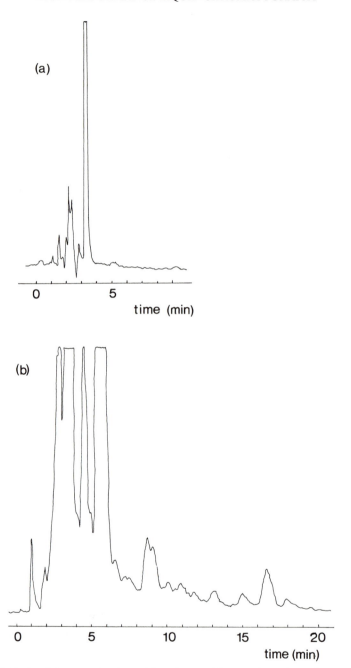

Figure 7.9 Study of long-running impurities of an anti-leprosy drug. Same detector attenuation for (a) 0.1 mg ml^{-1} solution; (b) 1.0 mg ml^{-1} solution. Better signal to noise in situation (b); however estimation of unknown impurities by calculating % of total peak area is more inappropriate than situation (a) since the main peak may well lie outside the linear range.

where A_1 is the peak area of the compound peak in the reference solution chromatogram, A_2 is the peak area of the compound peak in the sample analytical solution chromatogram, c_1 is the concentration of the compound in the reference solution and f is the concentration (or dilution) factor.

As described for purity determinations, care must be taken over the sequencing of samples and standards.

If little is known about the expected concentration of the compound and the concentrations at which the compound is present is likely to show great variations, it would be more appropriate to use standard solutions to create a calibration plot rather than to use the equation shown above.

Bibliography

Berridge, J. C. (1985) *Techniques for the Automated Optimisations of HPLC Separations*, Wiley, Chichester.

Miller, J. C. and Miller, J. N. (1988) *Statistics for Analytical Chemistry*, 2nd edition, Ellis Horwood, Chichester.

Wright, A. (1990) Strategies for mobile phase optimisation in HPLC, *Chromatogr. Anal.*, 5–7 April.

8 Sample preparation

D. STEVENSON

8.1 Introduction

For many analytical methods using HPLC as the end step, the samples presented to the analyst cannot be injected directly into the instrument. For example, the sample may be a large volume of water where the analytes are present at low concentrations, a solid such as soil or food-stuffs, or a biological specimen where numerous other compounds are present. Sample preparation is therefore needed to isolate the analytes of interest, to pre-concentrate them in order to lower detection levels and also to protect the analytical column from substances which may potentially damage the bed of packing material. Despite many advances in instrumentation for HPLC, only rarely are samples (particularly complex samples) injected without some form of sample pre-treatment.

The aim of sample preparation is to produce the sample in a form suitable for introduction into the measuring instrument. In the case of HPLC, where the mobile phase is liquid, the sample should ideally be presented dissolved in the mobile phase. If this is not feasible, then it should be dissolved in a liquid that is chemically very similar to the mobile phase, or at the very least a liquid compatible with the mobile phase. Hence most sample preparation procedures, irrespective of the original matrix, are aimed at extracting the analytes into a liquid. The isolation of analytes from a complex matrix is, in many cases, the rate-limiting step in an analytical procedure as well as the source of major errors. Sample preparation should, however, be considered an integral part of a whole analytical procedure. Many different steps have been used for the preparation of samples for HPLC. These are summarised in Table 8.1. This chapter concentrates on the procedures most commonly used to treat samples prior to HPLC.

8.2 Sample preparation procedures

8.2.1 Solvent extraction

The most common form of sample preparation is the extraction of analytes from one liquid phase to another. Clearly for this type of procedure

Table 8.1 Sample preparation steps prior to HPLC

Liquid handling	Centrifugation
Solvent extraction	Evaporation
pH change	Freeze drying
Chemical derivatisation	Enzyme digestion
Solid phase extraction	Hydrolysis
Söxhlet extraction	Column switching
Steam distillation	Low temperature storage
Supercritical fluid extraction	Filtration
Homogenisation	Column chromatography
Precipitation	Microwave irradiation
Dialysis	

the two liquid phases must be immiscible, and it is more common to be extracting analytes from an aqueous sample such as drinking water or a body fluid (for example plasma, serum or urine) into an organic solvent. If the partition of an analyte, A, between two liquids is represented by

$$[A]_{aq} \rightleftharpoons [A]_{org} \qquad (8.1)$$

then the partition coefficient, K, is the equilibrium constant for the reversible equilibrium

$$K = \frac{[A]_{org}}{[A]_{aq}} \qquad (8.2)$$

where $[A]_{aq}$ is the concentration of analyte in the aqueous phase and $[A]_{org}$ is the concentration of analyte in the organic phase.

Hence if the partition coefficient of an analyte between dichloromethane and water is four, then after extracting an aqueous solution with an equal volume of dichloromethane, four times the concentration of analyte would be in the dichloromethane as in the aqueous layer. The fraction of analyte extracted into the organic solvent is also dependent on the volume of the respective phases, hence increasing the volume of the organic layer would increase the fraction of analyte extracted (and therefore the recovery).

Table 8.2 Solvent polarity

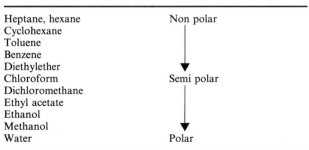

Heptane, hexane	Non polar
Cyclohexane	
Toluene	
Benzene	
Diethylether	
Chloroform	Semi polar
Dichloromethane	
Ethyl acetate	
Ethanol	
Methanol	
Water	Polar

Analytes are extracted into organic solvents only if they possess sufficient lipophilic character to be attracted into a non-polar solvent. When planning a solvent extraction strategy, it is necessary to consider solvent polarity. A range of commonly used organic solvents is shown in Table 8.2. When choosing a solvent to extract a solute from a complex matrix the aim is to achieve a high recovery of analyte, but also to leave behind as many potentially interfering compounds as possible. In order to achieve this, the approach is often to use the least polar solvent that will still give good recovery. Other factors affecting the choice of solvent for extraction include: its boiling point, since it is often wished to evaporate the solvent at a later stage of the method; its density, since it is sometimes wished to remove the solvent as the top layer (most organic solvents) or occasionally have it as the bottom layer (chlorinated solvents); toxicity; and purity.

For analytes that are ionisable, the other factor to consider in solvent extraction is pH. Using a compound with a carboxylic acid function (RCOOH) or a base (B) as an example, then the compound can exist in its ionised or its unionised form as shown below.

The form that predominates will depend on the pK_a or pK_b of the compound and the pH. The unionised form will be more soluble in the organic phase and the ionised form more soluble in the aqueous phase.

$$
\begin{array}{cc}
\underset{\text{unionised}}{\text{RCOOH}} \rightleftharpoons \underset{\text{ionised}}{\text{RCOO}^- + \text{H}^+} \\[2ex]
\underset{\text{ionised}}{\text{BH}^+} \rightleftharpoons \underset{\text{unionised}}{\text{B} + \text{H}^+}
\end{array}
\qquad (8.3)
$$

$$\text{acidic} \longleftrightarrow \text{basic}$$
$$\text{pH}$$

Hence an acidic pH will optimise the extraction of an acid into organic solvent and a basic pH will optimise the extraction of a base. This effect is known as ion suppression.

The optimisation of conditions for solvent extraction thus requires selection of solvent polarity and pH for ionisable compounds. One further approach for strong acids or bases (which are often difficult to extract quantitatively using ion suppression) is to use ion-pair techniques. With ion pairing the pH is selected in order to promote ionisation and a pairing ion of the opposite charge to the analyte is added. This forms a less polar, electrically neutral complex with the analyte, which aids extraction into an organic solvent. Ion pairing reagents often contain a hydrocarbon chain or other lipophilic moiety to enhance the extraction into organic solvent; for example the sodium salts of pentane, hexane, heptane, or octane sulphonic acid are often used to extract bases. Quaternary ammonium compounds (for example tetrabutylammonium phosphate) are the most common pairing ions for the extraction of acids. The simplest equation to describe ion pairing is shown below:

$$B^+_{aq} + P^-_{aq} \rightleftharpoons BP_{org} \qquad (8.4)$$

where B is the analyte which is basic and P is the pairing-ion. The equilibrium constant for the reaction is given by

$$K = \frac{[BP_{org}]}{[B^+_{aq}][P^-_{aq}]} \qquad (8.5)$$

The distribution of B between the phases can be represented by

$$\frac{[BP_{org}]}{[B^+_{aq}]} = K[P^-_{aq}] \qquad (8.6)$$

Thus the concentration of pairing ion can regulate the extraction of B into the organic phase. Ion-pair extraction is therefore a useful option when trying to extract strong acids or bases.

8.2.2 Solid-phase extraction

An alternative to liquid–liquid extraction is solid-phase extraction (SPE). With SPE a liquid sample is introduced into the top of a plastic syringe shape column containing a small amount (often 100–500 mg) of a selective adsorbent (Figure 8.1). The adsorbents are of the same types as used for HPLC, typically silica, or bonded silica such as C18, C8, C5, C2, cyano, phenyl, diol and ion-exchange materials. The properties of the adsorbents are similar to HPLC columns and so the same principles apply for retention and desorption of analytes.

The aim with SPE is to selectively isolate solutes from a liquid sample. The matrix is applied at the top of the column and pulled through the

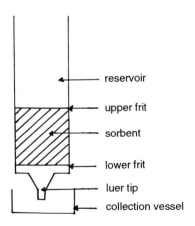

reservoir

upper frit

sorbent

lower frit

luer tip

collection vessel

Figure 8.1 A typical solid phase extraction cartridge. The sorbent is usually contained in a plastic syringe shaped column.

Table 8.3 Solid phase extraction cartridges

Type	Retention
C18 silica	Hydrophobic
C8 silica	Hydrophobic
C2 silica	Hydrophobic
Cyclohexyl	Hydrophobic
Phenyl	Hydrophobic
Cyanopropyl	Hydrophobic
Diol	Polar groups
Silica	Polar groups
Aminopropyl	Polar groups
Benzene sulphonic acid	Cation exchange
Carboxylic acid	Cation exchange
Quaternary amine	Anion exchange
Aminopropyl	Anion exchange

adsorbent bed under a vacuum (occasionally positive pressure is used). The analytes of interest are retained while many interfering compounds are not. The analytes are then eluted from the SPE column in a small volume of a suitable elution solvent. With some methods, one or more wash steps are introduced. The SPE cartridges normally require some priming before use, typically a wash with $1 \, cm^3$ of both methanol and a buffer. As with HPLC mobile phases, there are many combinations of eluting and washing solvent for each column type. The most commonly used phases are C18 and C8 silica where analytes are retained on the basis of hydrophobic interactions. The ability to elute the analytes in a small volume allows the concentration of a large volume of sample (e.g. water) to a much smaller volume without the need to further reduce liquid volumes. The 'trace enrichment' can allow the detection of much lower levels of analytes. Solid-phase extraction materials are in many cases replacing liquid–liquid methods as they are more easily automated. Table 8.3 shows some of the phases most commonly used for sample preparation.

8.2.3 Söxhlet extraction

With solid materials (such as foods) the initial step in sample preparation involves Söxhlet extraction. With this technique a solid sample is placed in a permeable thimble above a flask of solvent but below a cold water condenser (Figure 8.2). The extraction solvent is boiled, condenses and runs down into the part of the apparatus containing the sample. As the container fills, the sample is extracted with solvent. The apparatus is designed such that as the solvent level reaches a certain height it siphons back into the flask at the bottom along with any extracted residue. The process is then continuously repeated often for many hours. The end result is a

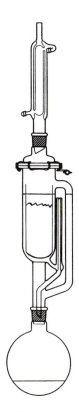

Figure 8.2 Söxhlet extraction apparatus.

fairly large volume of volatile liquid which is often further processed before analysis. It is often used in procedures to determine residue of pesticides in soils and foodstuffs. Modern apparatus allows the option of lowering the sample into the boiling solvent.

8.2.4 Protein precipitation and enzyme hydrolysis

Many HPLC analyses are carried out on biological samples such as blood, plasma, serum or urine. If urine analysis is to be carried out, the compound of interest is often excreted as the glucuronide or sulphate conjugate. These are very polar and are rarely determined directly. Samples are usually hydrolysed by acid, or an enzyme preparation to release the parent compound before analysis proceeds.

Blood, plasma, serum or saliva are rarely determined directly by HPLC. Proteins and fats present in such samples are known to have a deleterious effect on HPLC columns. The minimum pretreatment for such samples is to denature (and precipitate) the proteins present. This can be achieved by a variety of methods such as heating or treatment with acids, bases, water-miscible organic solvents such as methanol or acetonitrile or by the addition of chaotropic reagents. These methods will cause the proteins to precipitate such that the supernatant can be removed after centrifugation. Whichever method is used, care must be taken that the analytes are not trapped by denatured protein, resulting in low recovery.

8.2.5   Column switching

Column switching in HPLC offers an alternative for the processing of liquid samples such as biological fluids or water. Samples are injected directly onto one HPLC column and the analytes retained. After a set time a second solvent elutes the analytes onto a second analytical column for separation and detection. There are several alternatives in column switching such as the 'back-flushing' of retained components. It is also possible to use more than two columns. Columns can be of the same type or contain different stationary phases. The valves involved in column switching are automated so the technique offers the possibility of complete automation.

8.2.6   Dialysis

Dialysis involves the separation of higher molecular weight compounds from those of lower molecular weight by virtue of their differing permeation rates through membranes. The analyte solution is called the donor solution and the solution into which the analyte flows is the recipient. The main disadvantages of the technique are that the diffusion process across the membrane is relatively slow and the analyte is transferred to a dilute recipient solution. The greater the concentration difference between the donor and recipient solutions the faster the analytes diffuse. The most common use of dialysis is in the ASTED system where the recipient solution is constantly replaced in a flowing system. This dilutes the solution but in the ASTED system the dilute solution is concentrated on-line using an extraction cartridge. After enrichment the cartridge is switched in-line to an HPLC system for analysis. The method is thus completely automated and its use is growing particularly for the analysis of trace organics in biological fluids.

8.2.7   Techniques for the reduction of liquid volumes

In many analytical methods there is the need to reduce liquid volumes in order to allow determination of lower limits of detection. Solvent

Figure 8.3 Kuderna-Danish concentrator.

evaporation is also carried out in order to redissolve analyte in a different solvent, e.g. HPLC mobile phase. Solvent reduction can be achieved in a number of ways but care must be taken that losses of analyte do not occur. With small (say $10\,cm^3$ or less) volumes of volatile organic solvents, a favoured method is to blow nitrogen across the surface of the liquid while gently heating the sample container in a heating block or water bath. For volatile or unstable analytes the procedure can be carried out at room temperature or below.

Large volumes of solvent or indeed aqueous samples can be dried in a rotary evaporator operated under vacuum or at reduced pressure. The liquid evaporated is condensed into a reservoir. An alternative is to use a Kuderna-Danish concentrator which can also reduce several hundreds of millilitres of solvent to a few millilitres. The apparatus is shown in Figure 8.3.

8.2.8   Digestion of inorganics

Although mostly used for the analysis of organic compounds, HPLC is also used for the determination of inorganic compounds such as metals and anions using 'ion chromatography'. The analysis of inorganics often involves an initial digestion step using strong acid or alkali, particularly from biological or solid matrixes.

8.3   Matrix properties

When considering sample preparation step(s) the analyst has to consider the properties of the sample matrix as well as those of the analyte. Although many different types of analysis can be carried out on the same matrix some general comments can be made about a particular matrix.

8.3.1   Water

Concern about possible risk to human health posed by trace levels of toxic compounds has necessitated routine monitoring of drinking water quality. Waste water, water entering treatment plants, and groundwater are regularly assayed for trace organics such as pesticides, polyaromatic hydrocarbons, phenols and other trace organics. With this type of analysis there is a large volume of sample available and numerous compounds present mostly at very low concentrations. Water is, however, considered a relatively clean matrix. The usual requirement for sample preparation is to isolate the compounds of interest from the large volume of water. This is most commonly achieved using liquid–liquid extraction using a large volume of water extracted into a smaller volume of organic solvent. Often two or three solvent extractions are combined in order to achieve maximum recovery. Subsequent analysis then requires evaporation of solvent (hence a volatile solvent is chosen to speed up this step). The other common approach is to use solid-phase extraction. Several hundred cubic centimetres of water can be passed through a cartridge while trace organics are retained on the solid phase. Many organics can then be eluted in as little as $0.5\,cm^3$ of a suitable organic solvent, thus giving a 200-fold concentration from $100\,cm^3$ of water.

8.3.2   Biological fluids

The analysis of drugs and metabolites in biological fluids, particularly plasma or serum is one of the most demanding but one of the most common uses of HPLC. Blood, plasma or serum contain numerous endogenous compounds often present at concentrations much greater than those of the analyte. Analyte concentrations are often low, and in the case of drugs, endogenous compounds are sometime structurally very similar to the drug to be measured. One of the most important features of blood, plasma or serum analysis is the presence of proteins at relatively high concentrations. The binding of analytes to the plasma protein is usually broken down by the protein precipitation methods described earlier. The supernatant after centrifugation is thus protein-free.

Liquid–liquid or solid phase extraction are also very commonly carried out on plasma or serum, either directly or on a protein-free solution. A single solvent extraction of plasma may not provide a clean enough extract for trace analysis. An extra step whereby charged analytes are 'back extracted' into an acidic or basic aqueous solution is one means of further cleaning up in the sample. In some methods the pH of the aqueous solution is then altered again and subsequent re-extraction takes place. One other feature of blood, plasma or urine is that only low volumes are available for analysis.

The other biological fluid commonly assayed is urine. Unlike plasma or serum this contains only low concentrations of protein so protein removal is not necessary. The composition of urine varies widely and so does the volume passed by different individuals. Generally speaking large volumes are available for analysis. One of the major problems with urine is the large number of small organic compounds that are present. Liquid–liquid extraction (often preceded by conjugate hydrolysis) is often the sample preparation procedure used for urine analysis.

Tissue samples such as (liver, kidney, fat) share many of the features of blood, plasma or serum. However, they require an initial homogenisation step to attempt to solubilise the analytes in a suitable reagent, e.g. a buffer. Subsequent sample preparation then has much in common with blood, plasma or urine. The initial homogenisation is carried out by mincing using a sharp, rotating blade. Sometimes salt is added to disrupt the cellular material. The resulting homogenate, often in the form of a slurry, can then be centrifuged and the supernatant taken for further sample preparation. One of the practical problems associated with tissue analysis is that checking analyte recovery is difficult. Although recovery experiments can be carried out by spiking the analyte into a homogenate, this is not spiking into the actual matrix (cf. plasma or water samples where it is possible to spike into the actual liquid matrix).

There are several other biological samples less commonly analysed; these include bile, sweat, saliva, faeces, lung and bone. Of these, saliva analysis has gained in popularity, chiefly because this fluid is easy to collect. Volumes are low, however. The concentration of analyte in saliva is often quite close to that in plasma (although the pH difference between saliva and plasma means that for some ionised compounds it is significantly different). From the viewpoint of sample preparation, saliva can be thought of as a filtered blood, i.e. protein and lipid concentration is much lower, so sample preparation is similar to plasma or serum but does not require the initial protein precipitation step.

8.3.3   Milk

The determination of trace organics in both cow's milk (to check for impurities such as antibiotics) and human milk (to check for exposure of infants to toxic chemicals such as pesticides) has led to an increase in the number of assays carried out on this matrix. Milk has many of the features of plasma and serum in that fairly low volumes are available (from humans), there is the possibility of numerous small molecular weight compounds that might interfere with analysis, and analytes are present at low concentrations. Milk, however, contains much more lipophilic material than plasma or serum and this considerably affects the choice of sample preparation procedures. Because of the fatty nature of the milk, lipophilic contaminants tend to concentrate in this matrix. Many methods for the analysis of organics in milk therefore begin with a solvent extraction step using a non-polar solvent, such as hexane. This aims to extract total fat such that trace organics are fully extracted. This does mean that further clean-up such as solid phase extraction or back extraction is necessary. For many assays a proportion of the total fat extract is evaporated to dryness and weighed such that the concentration of analyte can be referenced per cubic centimetre of milk or per gram of fat. One particular problem associated with milk is the formation of cream layers at the interface between the layers when carrying out a liquid–liquid extraction.

8.2.4   Soil

Soil samples are often analysed for residues of pesticides and their breakdown products. Concern about the environment means that lower and lower limits of detection are required, and multi-stage sample preparation procedures are required. Soil samples are collected in the field so it is important that a representative sample is taken, as a small sample might not be representative. Samples are usually stored deep frozen until analysis in order to prevent decomposition on storage. Stones and residual plant material need to be physically removed and lumps broken up. The soil sample is then extracted with solvent using the Söxhlet apparatus shown earlier or sometimes by adding the solvent and sample to a jar and tumbling or homogenising. The extracts are then evaporated to a low volume or dryness and further cleaned up by SPE (either in a small cartridge or using a glass column) and by liquid–liquid extraction. Multi-stage methods are thus very common for soil samples.

8.3.5   Crops and food

Many different types of fruit and vegetable crops and other processed foods are analysed for a variety of trace organics both for impurities (such as pesticides) and for content (e.g. amino acids, lipids, etc.). As with soil, it is important to take a representative sample for analysis.

Most soft materials such as tomatoes, grapes or bananas are homogenised and then solvent extracted before further sample processing is carried out. However, material such as potatoes or apples have to be chopped first and the initial step can be Söxhlet extraction or homogenisation. Some materials (for example oranges or potatoes) may have to be peeled depending on whether the aim is to analyse the peel, the inner part of the food or the whole sample.

The nature of the matrix influences the initial extraction step and choice of solvent. Crops are sometimes classified on the basis of aqueous, dry and oily. For example, samples of low water content might be extracted with a mixture of acetone (miscible with water) and hexane. Samples with a high fat content might be extracted with a non-polar solvent such as hexane. As with soil samples, further clean-up by liquid–liquid partition and SPE or column chromatography is often necessary.

8.3.6   Air

Growing concern about occupational and environmental exposure to toxic substances has led to an increase in the amount of air analysis carried out. As a matrix, air is relatively clean and there is a large volume available. Although potentially there is the possibility of many substances being present, in practice their numbers are usually limited. The major problem with air is therefore the initial sampling step. One of the most extensively used methods for sampling air is to pull air through a solid adsorbent tube at a known flow rate for a set time. Trace organics in air will be adsorbed onto the sampling tube. Alternatively, air can be pulled through a trapping liquid in which the analyte is soluble or indeed a chemical reaction occurs with a trapping reagent. In the case of an adsorbent tube, sample preparation subsequently involves desorbing the analyte molecules using an organic solvent and hence a liquid sample is produced. Aerosols (suspensions of solid or liquid particulates in air) are collected onto a filter. Trapped analytes are then dissolved in a suitable solvent. After the initial trapping or trapping/desorption step samples are often injected directly into the HPLC. If further sample preparation is required, liquid–liquid extraction or SPE is the most common approach.

8.4   Analyte properties

The properties of the compounds to be assayed can help with the choice of sample preparation procedures. Hydrophobic compounds (containing few or no polar groups) are easily extracted into organic solvents or are easy to retain using hydrophobic attractions on C18 or C8 bonded silica cartridges. Semi-polar compounds require a more polar solvent mixture to extract them. Compounds containing polar functional groups such as hydroxy, amino, or carboxylic acid groups may be difficult to extract into organic solvents but can be selectively retained by solid adsorbents or polar bonded phases in SPE cartridges or columns. If compounds contain acidic or basic functional groups then the optimum pH must be chosen for solvent extraction (i.e. choose a pH to suppress ionisation). If the compound is strongly acidic or basic, ion-pair extraction may offer an alternative.

If the analyte has a high vapour pressure at room temperature then reduction of liquid volumes is more difficult, as losses of analyte may occur when evaporating the solvent. For volatile compounds, gas chromatography rather than HPLC is normally the method of choice.

Analytes containing functional groups that give a good response to one of the selective detectors may also require less sample preparation, particularly trace enrichment. For example compounds containing highly conjugated aromatic rings will respond to the fluorescence detector and those with easily oxidisable or reducible groups to the electrochemical detector. The extra sensitivity and selectivity provided by the detector means that less sample preparation is needed.

8.5   Examples of applications

Some examples of common methods are detailed below along with general notes on the choice of the sample preparation procedure and the factors influencing the choice.

8.5.1   Tamoxifen in plasma

The requirement is for a method capable of measuring this anti-cancer drug down to 20 ng/ml. The structure is shown in Figure 8.4a. The drug is lipophilic so solvent extraction is possible. The basic functional group means that high pH is necessary for extraction. Endogenous compounds such as proteins and fats would cause rapid deterioration of HPLC columns if the sample were injected directly. The method involves raising the pH of the plasma ($1 \, cm^3$) using a small volume of ammonium hydroxide and then extracting with diethyl ether. The ether layer is then

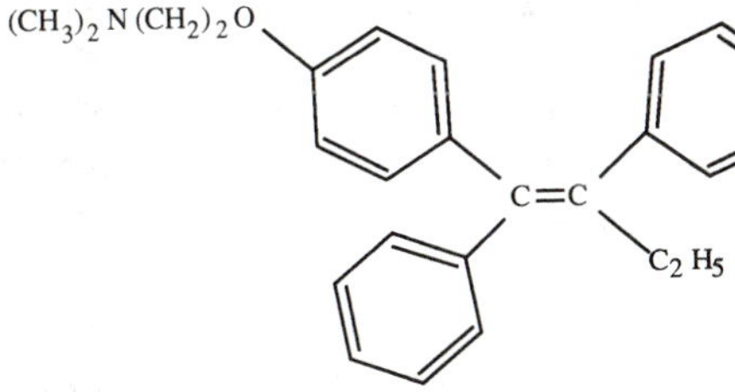

(a)

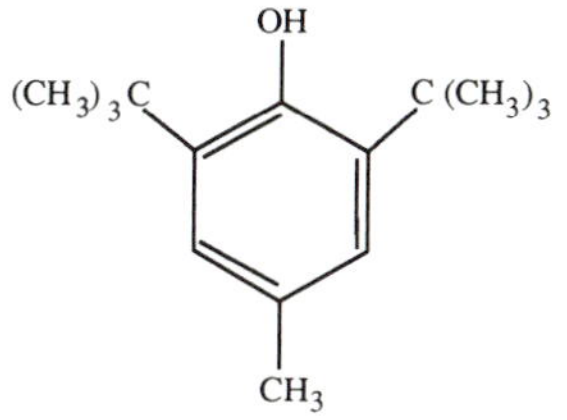

(b)

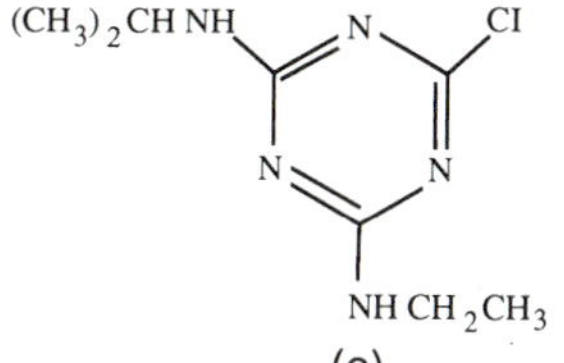

(c)

Figure 8.4 Chemical structures: (a) tamoxifen; (b) butylated hydroxytoluene; (c) atrazine.

evaporated to dryness under nitrogen (very easily done as the solvent is volatile). The extracts are then reconstituted in 0.1 ml of mobile phase so that a tenfold concentration factor is achieved.

8.5.2   Butylated hydroxytoluene in animal diet

The requirement is a method capable of measuring this anti-oxidant down to 250 mg/kg. The compound (Figure 8.3b) contains an aromatic group and is suitable for HPLC with UV detection. The solid matrix requires some kind of extraction to release the analyte into a liquid (preferably volatile) phase. The concentrations required are quite high but the matrix will contain many fatty compounds that could cause the HPLC column to deteriorate rapidly. The method involves taking 10 g of sample, adding 100 cm^3 of methanol (in which the analyte is very soluble due to the hydroxy group) and homogenising for 1 min. An aliquot of the supernatant is then taken and evaporated to dryness under nitrogen and redissolved in HPLC mobile phase. This very simple sample preparation procedure is possible only because of the high concentrations involved. If a lower limit of detection were required, the probable approach would be to use a Söxhlet extraction to ensure quantitative recovery of analyte at the very low levels. This would co-extract a lot of other material and so further clean-up using liquid–liquid extraction and SPE would be required.

8.5.3   Pesticides in water

Pesticides such as atrazine (Figure 7.4c) are determined in water down to the EC drinking water limit of 100 ng/l. Atrazine is a basic compound so liquid–liquid extraction at high pH is possible. However, solid phase extraction using C18 or C8 bonded silica is now preferred. Typically, the cartridge is 'primed' by passing through a small volume of methanol, the sample (100 cm^3) is loaded and analyte retained due to the hydrophobic nature of the cartridge and the aromatic ring on the analyte. The analyte is then eluted from the cartridge by a small volume (say 1 cm^3) of a semi-polar solvent such as ethyl acetate, and injected directly onto the HPLC. A 100-fold concentration factor is then achieved.

8.6   Automation

As stated earlier, the rate-limiting step of many assays is the sample preparation. Not surprisingly, much effort has been devoted to attempts to automate sample preparation where procedures are lengthy and where many samples are analysed. Successful automation can lead to a greater throughput of samples, more accurate methods, cost savings and can free laboratory staff to undertake more interesting and productive work. Automation of instrumentation and data handling is much easier than automation of sample preparation since instruments and data are used for many different types of analysis. Sample preparation procedures depend on both the matrix and the analyte properties and are therefore rather more application specific.

Solid phase extraction systems have been successfully automated (for example in the ASPEC system, marketed by Gilson). This allows samples to be applied to the cartridge and is linked on-line to the HPLC. An extension of this is the ASTED which combines dialysis with trace enrichment on a solid phase and is automatically coupled to the HPLC. These methods have found widespread application for the analysis of drugs and metabolites in body fluids. The fact that solid-phase extraction methods may potentially be automated is one further reason for their current popularity in the manual mode; if many samples are expected the manual method can be automated. This is much more difficult with liquid–liquid extraction.

Another approach to automation has been the development of robots or more correctly robotic arms. These are reprogrammable mechanical arms which can carry out most of the tasks that humans carry out, but of course can handle toxic substances and can work longer hours in a continuous period. Most robotic configurations consist of the control arm with a series of workstations in a circle around the processing arm. The arm moves from one workstation to the next carrying out its sampling tasks. Adding reagents, solvent extraction, centrifugation, phase separation and transfer to HPLC autosampler have all been proven possible.

As more and more samples are sent for analysis, automated procedures will be required. Improvements in instrumentation and software will facilitate the development of greater automation of sample preparation into a fully integrated analytical method.

8.7 Concluding remarks

The extent of sample preparation required for a particular analytical task will depend on a number of factors. It must be remembered that the analyst really does not want to do any sample preparation at all, so efforts are devoted to carrying out the minimum amount of sample preparation for a particular assay. The following factors influence the amount and choice of sample preparation:

- the chemical nature of the analyte
- the complexity of the matrix
- the likelihood of interfering compounds
- the physical nature of the matrix
- the amount of sample available
- the specificity of the end step (by detector)
- the detection limit required
- the chemical stability of the analyte
- the accuracy required

Sample preparation strategies therefore tend to be very application orientated. When developing methods, it is advantageous to work in a logical and stepwise manner but recognising that potential problems are interlinked. As with other chromatographic methods, the use of internal standards closely related chemically to the analyte(s) of interest are extremely valuable. These should be added as early as possible in the procedure, for example, spiked directly into a plasma or water sample. Whenever possible, calibration standards should be added directly to the matrix and these are then processed exactly as for samples containing unknown levels of analyte.

When developing a method it is necessary to obtain satisfactory chromatographic conditions for pure standards such that sample preparation steps can be evaluated. When testing certain procedures it is necessary to consider both the recovery of analyte and to check for compounds from the matrix that might have the same chromatographic retention as the analyte (or internal standard). Hence both spiked samples and blank are tested during method development. The availability of radiolabelled analyte can prove extremely valuable when carrying out method development. If good recovery of analyte and a clean blank is obtained the analyst will usually carry out a method validation programme. The most important feature of this is the reproducibility. If a method is not satisfactory then it is necessary to investigate which step in the procedure is causing the problem. In this situation it is sensible to spike analyte at various stages of the method and process samples. For example, a typical method might involve taking $1 \, cm^3$ of plasma, adding internal standard, making alkaline and then carrying out a liquid–liquid extraction with diethyl ether, mix, separate layers, evaporate to dryness, redissolve in mobile phase and inject onto the HPLC. If this method gave irreproducible recovery it would be informative to take some plasma, extract and add analyte to the solvent extracts and process. This would reveal whether the actual extraction from plasma caused the problem. If not the already dried down extracts of blank plasma could be spiked. A logical stepwise investigation should pinpoint the stage of the procedure that is giving the problem such that a suitable modification can be made.

There are many potential pitfalls with multistage procedures, but the need for lower limits of detection and the requirement for more analyses means that sample preparation is a growing need for many HPLC laboratories. Problems might include: instability of analyte on storage or during the sample preparation procedure; interferences from the matrix, reagents or containers in which samples are stored; incomplete release from the matrix; emulsion formation when extracting; adsorption of traces of analyte on to drying down vessels; or low recovery due to the choice of an unsuitable procedure.

Many advances have been made in recent years in chromatographic instrumentation and data handling including important advances in automation. However, for many methods, equally important is the sample preparation procedure. These are sometimes complex, and are often the key step in a method, but perhaps receive less attention than is warranted. The optimisation of sample preparation and the development of new procedures will provide analysts with a difficult and challenging problem.

Bibliography

Chamberlain, J. (1985) *Analysis of Drugs in Biological Fluids*, CRC Press, Boca Raton, FL.
Beyermann, K. (1984) *Organic Trace Analysis*, Ellis Horwood, Chichester.
Reid, E. (1981) *Trace Organic Sample Handling*, Ellis Horwood, Chichester.

9 Polymer analysis

A.J. HANDLEY

9.1 Introduction

Industrial or synthetic polymers find extensive use in modern day society. They are many in number, polyvinyl chloride (PVC) polyamides (Nylon), polyethylene teraphthalate (PET), polystyrene and polyolefins, to name but a few. Polymers are without exception very complex compounds, capable of manifesting themselves in many shapes and forms. They can exist as viscous liquids, powders, coloured granules, cast or extruded sheet, transparent or translucent film, formulated (in some cases in excess of ten different additives may be added) or unformulated. Hence they can present a very daunting task to the analyst or polymer chemist wishing to fully characterise such materials.

Polymers are described as polydispersed compounds, that is they consist of a mixture of molecules with the same chemical composition, but with various molecular sizes. Copolymer and terpolymer systems can also vary in chemical composition. In addition they may contain monodispersed components such as polymer additives and monomers.

The type of analytical determinations in this area are in many cases unique. Broadly speaking they can be split into three areas:

(i) the characterisation of the polymer in terms of its molecular weight/ mass distribution, and where possible the production of absolute molecular weight averages;
(ii) gaining information on the chemical composition of the polymer;
(iii) identification/quantification of low molecular weight species in the polymers.

Given this, can liquid chromatography help in providing the necessary solution? Firstly, the chromatographic modes available must be considered.

9.2 Chromatographic modes used for polymer analysis

In HPLC, several modes of separation are available, adsorption, partition, ion exchange and size exclusion. The choice of separation mode depends largely on the nature of the solute to be separated. What therefore is the nature of a polymer? What are its physical parameters that will effect the separation process:

(i) A polymer can extend from 1000 molecular mass up to ten million and above.

(ii) Most polymers have a molecular mass range covering two to three decades.

(iii) Polymers in many cases contain low molecular weight species (<1000) such as monomers and additives.

(iv) Polymers are soluble in only a limited number of solvents, at usual HPLC operating temperatures.

(v) The diffusion coefficients of polymers in solution range between 10^{-6} and $10^{-9}\,cm^2\,s^{-1}$ which is some two orders of magnitude smaller than that of small molecule substances.

(vi) The pore sizes of conventional HPLC type column packing materials are comparatively small when compared to the sizes of polymers in solution.

These properties and effects make separation by the more conventional HPLC modes of partition/adsorption difficult for polymers, in comparison to lower molecular weight species.

In partition chromatography, for example, the mass transport processes are much slower than those for low molecular weight species, resulting in poorer and often non-reproducible separations.

In adsorption systems, the physical size of the polymer will prevents its access to many of the pores within the LC stationary phase thus considerably reducing the amount of surface available for interaction. The nature of a polymer's conformation in solution can also result in complex interactions on the surface of the stationary phase, often resulting in irreversible interaction.

In addition, the very action of injecting a polymer into a HPLC system can present problems for fragile high molecular size species as the strong irregular shear forces generated by the mobile phase flowing under pressure through a tightly packed chromatographic column bed, can result in the polymer physically shearing in its passage through the system.

Thus, interactive chromatographic modes (partition/adsorption) in general are not suited for the analysis of bulk polymers. However, they can be used for the analysis of the lower molecular weight species contained in polymer systems.

For polymers, a non-reactive, gentle separation mode needs to be provided; size exclusion offers the solution. This mode of HPLC as described later is the main method of separation used in the polymer industry. Hence the majority of this chapter focusses on size exclusion HPLC and its use.

9.3   Size exclusion chromatography

9.3.1   Introduction

Size exclusion chromatography (SEC) is a name that unites two of the more traditionally named technologies: gel filtration chromatography (GFC), used primarily with aqueous mobile phases and gel permeation chromatography (GPC), used with organic mobile phases.

Both techniques have their origins in the late 1950s, early 1960s, GFC from the work of J. Porath and P. Flodin and GPC (the more familiar name amongst polymer scientists) with the work of M.F. Vaughan and J.C. Moore. As its name suggests, SEC describes a technology where components are separated according to differences in size (hydrodynamic volume). The size in solution being a function of the sample's molecular weight, shape and degree of solvation. It is a highly predictable technique suitable for the separation of samples where the components are sufficiently different in molecular size, and as such the technology is ideally suited to the application of polymeric species.

9.3.2   Scope of the technique

The key features of SEC are as follows:

(i)   Used for the characterisation/separation of samples of molecular weight $< 10^2$ to 10^7.
(ii)   Can be used with both non-aqueous and aqueous mobile phases.
(iii)   SEC columns have limited peak capacity, thus complete resolution of complex examples is not normally achieved.
(iv)   For polymers, a two-fold difference in molecular weight is needed for significant separation.
(v)   For smaller molecules a weight difference of more than 10% is required for separation.
(vi)   Its primary use is in the polymer industry for establishing the molecular weight distribution of a sample, where individual components need not be resolved.
(vii)   In comparison to conventional HPLC, SEC is generally a more expensive technology often involving the use of dedicated chromatographs capable of operating at high temperature ($> 100°C$), with unique detectors and dedicated data handling systems.

9.3.3   Mechanism of separation

There are many theories which describe the mechanisms of SEC separation, a good review being provided by Janca (1984). The simplest way to

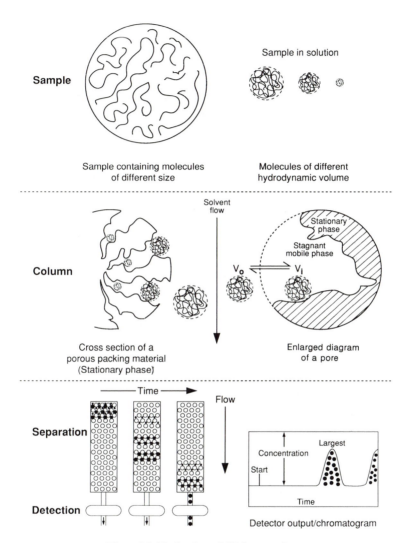

Figure 9.1 Mechanism of SEC separation.

describe the mechanism, is to adopt a pictorial approach (Figure 9.1). The sample or polymer is depicted as containing a number of coiled molecules (inherent of a size distribution). The individual molecules in solution will occupy different sizes depending on their hydrodynamic radius/volume.

The stationary phase is represented as a porous packing material containing different size pores (cross-linked inert gel). The mobile phase is normally a single solvent capable of dissolving the sample.

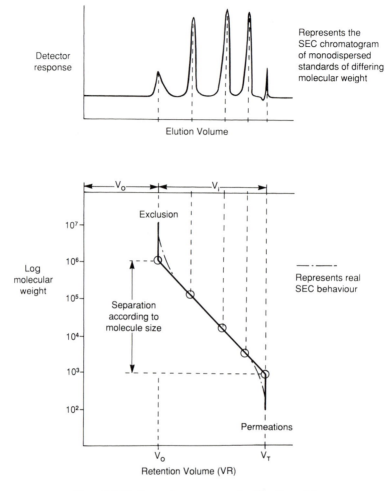

Figure 9.2 Elution volume curve and calibration plot.

As flow is applied to the system, solvent will flow only in the interstitial spaces around the gel particles as the channel produced between the stationary phase particles in the column is much larger than the pores inside the gels. These pores ultimately will contain stagnant mobile phase.

On injection of the sample molecules onto the column, molecules too large to penetrate and hence diffuse into the pores will be 'excluded' and elute from the column first. Medium size molecules may be excluded from the denser parts of the stationary phase pore structure, but diffuse freely in the larger passages of the pores 'selective permeation'. Molecules of comparable size to that of the solvent will distribute themselves throughout the entire pore structure and hence elute from the column last ('total permeation') (Figure 9.2). This model of the process is very much over-simplified, but serves as a useful basis for discussion in the following sections.

9.3.3.1 Retention in SEC. If we consider the separation of a number of monodispersed polystyrene standards of differing molecular size in a given SEC column packing. The retention of the standards can be represented by a graph of molecular size versus retention volume V_R (Figure 9.2).

V_0 is the exclusion limit of the column and equal to the interstitial volume of the packed SEC column (volume of eluent between the particles of stationary phase). V_I is the pore volume of the packing (the internal solvent volume in the pores). V_T is the total permeation volume. The volume of solvent at which a solute elutes from the column or the volume of liquid corresponding to the retention of a solute on a column is known as the retention volume (V_R). V_R can be related to the physical parameters of the column as follows:

$$V_R = V_0 + K_{sec}V_I \qquad (9.1)$$

where K is the solute distribution coefficient based upon the relative con-centrations between stationary and mobile phases; i.e.

$$K_{sec} = \frac{V_{I_{acc}}}{V_I} \qquad (9.2)$$

and $V_{I_{acc}}$ is the pore volume accessible to a given size of solution ($V_R - V_0$). V_I is the total pore volume ($V_T - V_0$), thus

$$K_{sec} = \frac{V_R - V_0}{V_T - V_0} \qquad (9.3)$$

Hence retention volume can be expressed in terms of two measurable quantities V_0 and V_T, and K_{sec} which is a function of solute molecular size and of the column packing pore size. In SEC, it can only vary between zero (total exclusion) and unity (total permeation).

As can be seen, SEC retention is considerably different in nature to that of other LC methods (different equations for both retention and distribu-tion coefficient). This is mainly due to SEC retention being an entropic rather than an enthalpic equilibrium process dependent on the shape and size of both the solute molecules and the stationary phase pores.

9.3.3.2 Resolution in SEC. The term resolution in this section refers to any factors affecting the SEC separation when applied to macromolecules. As the distribution coefficient K_{sec} is limited to the range 0 to 1, Giddings estimated that an SEC column could only accommodate a limited number of peaks (peak capacity), estimated as 20 analytes base to base. An SEC column will therefore never be able to fully resolve individual polymer homologues. How then do we maximise the resolution available?

SEC resolution (R_S) depends on both peak separation (ΔV_R) and column dispersion (δ_V) such that

$$R_S = \frac{\Delta V_R}{4\delta_V} \qquad (9.4)$$

Factors effecting the peak separation include column volume, particle porosity, pore size distribution and solute conformation. Column dispersion is influenced by column length, particle size and the mobile phase temperature, viscosity and flow rate.

In practice, SEC resolution can in the main only be increased by increasing the number of theoretical plates in the system. If SEC analysis is performed with a column of optimum pore size/pore distribution, the number of theoretical plates can be increased by:

(i)   increasing the column length, thus providing extra pore volume. Longer elution times will result; however, unlike normal LC, the longer retained peaks will have smaller peak dispersion; small molecules elute later in an SEC separation, these compounds have higher diffusion coefficients than the larger early eluting components, and hence are less susceptible to mass transfer zone spreading;

(ii)   reducing the particle size of the stationary phase, thereby increasing efficiency and making maximum use of the limited peak capacity;

(iii)   optimising flow rate; the efficiency of SEC columns can be very much affected by the mobile phase flow rate. For small molecules maximum efficiency is achieved at a rate of $1 \, \mathrm{ml \, min^{-1}}$. However as the molecular weight of the solute increases, an increase in efficiency is observed with decreasing flow rate, this can be attributed to the reduced mass transfer in and out of the pore matrix.

9.3.4   Experimental aspects of SEC

Size exclusion chromatography in the main uses conventional liquid chromatographic equipment. In most cases the polymer in solution can be injected directly onto the chromatograph, however in the case of formulated polymers which contain inorganic fillers and pigments some prior sample separation will be needed.

In the chromatograph, the sample flows through the SEC column or columns, the separation is effected and the eluted fractions of the sample or polymer are detected using a suitable concentration detector. An elution volume curve (chromatogram) is produced which is often then used for polymers to produce a distribution curve (section 9.3.5.1). An SEC system for polymer analysis (Figure 9.3) differs from a conventional HPLC system in:

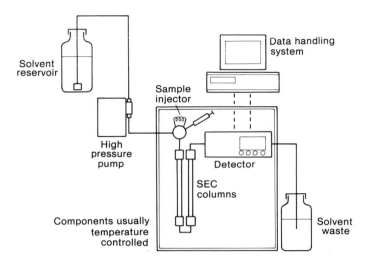

Figure 9.3 Schematic diagram of typical SEC apparatus.

(i)   in most cases a single solvent is used to elute the sample;
(ii)   the system is operated at elevated temperature (sometimes as high as 150°C) to ensure polymer dissolution;
(iii)   a series of columns are often used in the separation;
(iv)   unique detectors are employed to provide polymer related data;
(v)   multiple detection systems are used to obtain information on polymer molecular weight and composition;
(vi)   specialised data handling packages are needed to fully exploit the data produced from the SEC separation/detection systems.

9.3.4.1 Key components in SEC analysis

9.3.4.1.1 Mobile phase. The purpose of the mobile phase in SEC, is merely to act as a solvent for the sample. Hence it should be chosen to:

(i)   achieve optimum solubility of the sample;
(ii)   be compatible with the column packing used (gel-based SEC packing, being polymeric, will swell or shrink depending on the solvent used);
(iii)   permit detection.

For organic phase SEC, medium-polarity solvents such as tetrahydrofuran, toluene, chloroform and dichloromethane are used. However, more polar solvents such as dimethylformamide, N-methyl pyrrolidone or dimethylsulphoxide can be used, and for more specialised applications orthodichlorobenzene, trichlorobenzene and phenolic solvents at

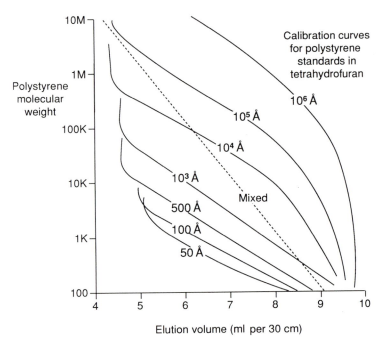

Figure 9.4 The size range of SEC columns. Reproduced with permission from Polymer Laboratories Limited.

temperatures in excess of 100°C are used (polyolefins, polyamides and polyesters). For aqueous SEC, water and buffer system are used. However care must be taken to avoid non-SEC separation mechanisms occurring.

9.3.4.1.2 The pumping system. Accuracy of mobile phase flow rate is very important in SEC, particularly when producing molecular weight distributions. This can be accommodated by most modern reciprocating pumps; however, the pump seals must be compatible with the solvent used in the separation. Typical flow rates of 1 ml min^{-1} are used.

9.3.4.1.3 The injector. HPLC type valve injectors are employed, but in some of the more specialised SEC systems, these are often designed for heating up to high temperature, to prevent dissolution of the polymer sample in the valve. In general injection volumes greater than 100 μl are used to increase detector response. Raising the polymer concentration should be avoided, as this can lead to an increase in the solution viscosity, causing problems with both sample injection and separation. Typical concentrations of sample are between 2 and 5 mg ml^{-1} depending on molecular weight.

9.3.4.1.4 The column. The heart of the SEC system is the column or column set. Most systems involving the use of organic or non-aqueous mobile phases, use polystyrene/divinylbenzene gels as packing. These materials are highly cross-linked, macroporous and spherical in nature. The divinylbenzene causes cross-linking in the polymer providing structural rigidity and spherical materials are produced when microspheres of the resin are fused together. The rate of fusion controlling the porosity of the final product. Particles of 3, 5 and 10 μm material can be produced.

These packings have an advantage over the more traditional silica SEC packings in that:

(i)   they have a wide range of solvent compatibility;
(ii)   they are easy to produce in a wide range of pore sizes $(50-10^6 \text{ Å})$.

For systems involving the use of aqueous solvents, silanised silicas and hydrophilic polymer based packings with either a polyhydroxyl or polyamide surface are used.

As stated previously, unlike conventional chromatography, SEC involves the use of columns coupled together in series. A number of differing pore size columns can be joined together to cover the desired sample molecular mass range. Figure 9.4 shows the range of columns available, the value in Ångstroms representing the mean pore width $(10 \text{ Å} = 1 \text{ nm})$.

Care should be taken, however, when linking different pore sizes together as occasionally artefacts in the separation can result. The above problem is more common when mixed gel packings (i.e. selected pore size gels are blended to produce columns of a wide linear molecular mass separation range) are used.

Columns are often linked together to increase resolution between components. In this case a single column might not possess a high enough resolution and hence a second column of similar exclusion characteristics can be added. This increase in column length leads to a proportionate increase in pore volume and hence separation power.

Modern SEC columns are capable of achieving efficiencies up to 80 000 plates per metre and are typically 300 mm × 7.5 mm. The strategy of using multiple columns ultimately results in longer analysis times than LC, an SEC separation normally taking between 20 and 30 min.

9.3.4.1.5 The detectors. As stated previously some unique detector systems are used in SEC to monitor the polymer separation. The nature of the mobile phase used, the operating temperature of the SEC and the sample itself can sometimes restrict the choice of concentration type detector used. In addition, polymers will usually be detected as distributions and not individual peaks. Because of this specific detector systems can be used to disseminate the data provided from the distributions, and to achieve molecular weight/mass information from the polymers.

Mass concentration detectors include the differential refractometer, ultraviolet and infrared detectors.

The differential refractometer (DRI) is the most widely used detector in SEC. It is considered a universal detector since most polymer solutions have a significantly different refractive index from the solvent itself. Two main designs of DRI are available, the reflectance or Fresnel and the deflection type. Both detect the change in refractive index of the solvent stream as the solute emerges from the column, with respect to a reference solvent stream. The signal can thus be both positive or negative. DRIs offer a wide range of linearity, can be used with both aqueous and non-aqueous mobile phases and can be operated at temperatures in excess of 100°C. The major weaknesses of the detector are:

(i)   DRI only offers modest sensitivity, hence is unsuitable for measuring low concentrations;

(ii)   they are very sensitive to changes in ambient temperature and pressure.

Ultraviolet absorption detectors (UV) are used in SEC when:

(i)   a complex mobile phase is used (aqueous SEC or organic phase SEC in a mixed solvent);

(ii)   the polymer or additive has a readily detectable UV chromatophore.

The most commonly used UV, as with HPLC, is the variable wavelength design. This detector uses the principles of the Beer–Lambert law. A deuterium lamp provides the energy source, a grating monochromator is used to select the specific wavelength for monitoring, the light is focussed onto a 10 µl flow cell through which the column eluent passes and the resulting energy is monitored by a photodiode. Wavelengths from 190 nm to 370 nm are available for detection, although some detectors will extend into the visible region. UV detectors are capable of very high sensitivity, good linearity and are relatively insensitive to mobile phase flow rate and temperature changes. The main limitations of UV detection in SEC are:

(i)   only a relatively small number of polymers exhibit an appreciable UV absorption;

(ii)   many of the commonly used SEC solvents are not suitable for use.

Infrared, like the UV and visible regions of the spectrum, provides a useful method for detecting solutes in liquid streams. The detector uses the principle of attenuated total reflectance, with a single beam spectrometer being used in conjunction with a low volume flow cell. The cell is equipped with sodium chloride, calcium fluoride or zinc selenide windows and can be heated to temperatures greater than 100°C. The detector can be used over the region of 2.5–14.5 µm and by selecting solvents that provide acceptable levels of transmission in appropriate spectral regions, it is possible to tune the IR detector for either universal or highly selective operation.

Like the UV detector, the IR detector is relatively insensitive to fluctuations around a set temperature and therefore particularly suitable for SEC at temperatures far above ambient conditions. One of the main uses for this type of detector is in the detection of polyolefins which can only be chromatographed at temperatures in excess of 135°C. Infrared detection is also useful in measuring polymer stoichiometry as a function of molecular weight/mass. Here the detector is used to monitor specific functional groups in a copolymer.

The detector is solvent limited in terms of its universal use, as a workable absorbance window has to be found to detect the polymer in the presence of a solvent, which itself can possess a strong infrared spectrum.

Molecular weight/mass detectors include light scattering detectors and viscosity detectors. When SEC is used in the characterisation of polymer systems, its main aim will be the production of a molecular mass/weight distribution and where possible absolute molecular weights. Mass calibration is a complicated matter (section 9.3.5.1) in that calibration curves differ for different polymer types, and for many commercial polymers, direct molar mass calibration is not possible because of the lack of suitable, known molecular weight standards.

An alternative is to determine the polymers molecular weight/mass in the SEC eluent *in situ*, by use of on-line molecular mass sensitive detectors. Two such detectors are commercially available, the light scattering detector and the viscosity detector. These detectors are usually used in series with a mass concentration detector and require specialised data handling/software to compute the outputs from the twin detectors and to produce molecular weight/masses and distributions.

Light scattering detectors, when used in conjunction with a refractive index detector, can provide absolute weight average molecular weights and distribution for homopolymers separated by SEC. The detector consists of a helium-neon laser operating on the 633 nm line. This wavelength provides a good signal-to-noise ratio even for weakly scattering solvents such as water. The light from the laser is passed through a series of attenuators then onto the detector cell consisting of two thick fused silica windows separated by a spacer and of 10 µl volume. Light scattered by the sample is collected and focussed onto a photomultiplier equipped with a digital readout. From the detector response the excess Rayleigh Factor can be calculated (Jordan, 1980) which is proportional to concentration and molecular weight. Hence by using a concentration detector in line with the light scattering detector to monitor the SEC separation, weight average molecular weight and molecular weight/mass distribution can be calculated.

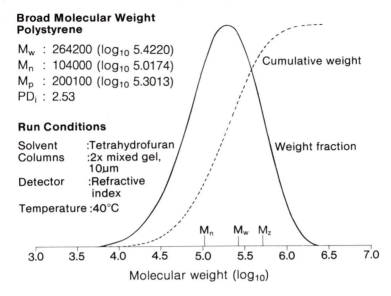

Figure 9.5 SEC molecular weight distribution curve.

The detector can be used in both aqueous and organic phase systems at temperatures ranging from 0 to 165°C. The detector suffers from lack of sensitivity to low molecular weight species. However as a stand-alone detector it offers a greater sensitivity for very high molecular species (greater than million molecular weight) than refractive index detection.

The viscosity detector, as with the light scattering detector, is used along with a refractive index detector, which is often housed within the same instrument to minimise the dead volume between detectors. The combined detector can provide absolute molecular weight (mass) averages, intrinsic viscosity and long-chain branching information for polymers (Kuo, 1987), when used to monitor the SEC separation. A number of commercial designs of viscometers are now available. All operate on the principle of Poiseuille's Law which states that the pressure drop across a capillary is proportional to the viscosity of the fluid and the flow rate.

Eluent from the SEC separation passes through a capillary or capillaries (one, two and four capillary designs are available) and the pressure drop is monitored by a differential pressure transducer. From the pressure drop and the concentration of the sample (obtained from the DRI) the intrinsic viscosity (η) can be calculated. Molecular weight averages, Mark-Houwink (for polymer branching studies) and molecular weight distribution plots can then be obtained after suitable data treatment.

The sensitivity of the viscometer is much greater for low molecular weight components than light scattering. For high molecular weight components, the converse is true. The detector can be operated with both aqueous and non-aqueous solvents and certain designs are capable of operation at temperatures up to 150°C. In all operations, precise flow rate and temperature control is needed to ensure detector baseline stability.

9.3.4.1.6 Data handling. As with all forms of HPLC equipment, data systems for chromatographic peak integration are used commonly in SEC. However, for polymer analysis, more comprehensive and specific software packages are needed to generate molecular weight distributions, weight averages, and evaluate data from dual detectors system. Thus in practice SEC can impose stringent requirements upon any data handling system.

9.3.5 *Applications in the polymer industry*

In many cases, the number of species occurring simultaneously in a synthetic polymer is so large, that it is not possible to separate them from one another. Hence the chromatogram obtained for a polymer by SEC seldom resembles the typical high efficiency separations obtained for small molecules by interactive chromatography (reversed-phase, etc.). However, the chromatogram of detector response versus retention volume, is a measure of the molecular size distribution, which can provide a vast amount of information on the polymer (Figure 9.5).

By suitable calibration of the resultant chromatogram a molecular weight distribution and associated molecular weight averages can be obtained. Subtle differences in the above parameters affect many of the end use properties of polymers such as:

- impact, tensile and adhesive strength;
- brittleness and drawability;
- cure time;
- melt and flow characteristics;
- solution properties;
- hardness.

The elution volume curve (chromatogram) or computed molecular weight distribution provides a qualitative fingerprint or picture of the polymer which can be used effectively to:

(i)   compare batch to batch variations in polymers;
(ii)   monitor polymer growth with reaction time;
(iii) monitor polymer degradation;
(iv)   look at the effect of different reactants (catalysts) on the polymer;
(v)   observe the disappearance/presence of low molecular weight compounds.

The chromatogram can also provide quantitative information on:

(i)   polymer/monomer ratios during polymerisation;
(ii) polymer blends;
(iii) levels of additives, monomers and initiators, providing good baseline separation can be achieved.

Hence SEC is an ideal tool for quality control in the polymer industry.

9.3.5.1 Determination of molecular mass/weight distributions and associated molecular weight averages by SEC. The production of molecular mass/weight distribution has been referred to in many sections in this chapter. How are they obtained? The output from the SEC system is an elution volume plot (concentration versus retention volume/time), with suitable calibration a molecular mass or weight distribution can be produced and the common molecular weight averages can be derived mathematically from the data (note: a polymer contains a distribution of molecular weights, hence its weight can only be defined as an average).

Common molecular weight averages calculated include: M_n, number average molecular weight; M_w, weight average molecular weight; M_z, Z average molecular weight; M_v, viscosity average molecular weight; and D, polydispersity $= M_w/M_n$.

9.3.5.1.1 Calibration of the SEC system. The two most common methods are:

(i) use of narrow dispersed polymer standards;
(ii) use of secondary standards.

The above methods are specific to SEC systems employing a single concentration detector, as has been noted in the previous section and some molecular weights can be obtained by the use of a dual detector system.

Narrow dispersed standards are the most commonly used. This involves the use of well characterised polymers standards of narrow molecular weight/size distribution and known molecular weight. A variety of polymer types are commercially available, polystyrene being the most widely used. These standards are run either individually or in well resolved combinations, alongside the polymer solutions under analysis. The elution volumes of the standards are then recorded and a calibration graph of log(molecular weight) versus elution volume constructed (similar to Figure 9.2) whose slope is equal to

$$\frac{\mathrm{d}\log M}{\mathrm{d}_v} \tag{9.5}$$

Having calibrated the system, the calculation of the molecular mass/weight distribution curve and molecular weight averages requires a number of data treatment steps (Evans, 1973).

(1) The sample chromatogram and those from the standards are assessed and any corrections for elution volume variation made.

(2) The start and end points of the polymer peak are defined and a baseline constructed on the chromatogram.

(3) The chromatogram is normalised to produce a plot of % dw/dv versus dv (dw = concentration, dv = elution volume).

(4) The dw/dv value is multiplied by the reciprocal of the slope of the calibration curve corresponding to the appropriate retention volume.

(5) This results in values of $dw/d \log M$ which are plotted against $\log M$ to yield the differential molecular weight/mass distribution curve (MWD).

(6) The individual molecular weight/sizes are then calculated from moments around the distribution.

It should be noted that as polymer size in solution is related to not only molecular weight, but also polymer shape or conformation, polymer standards of the same homologous series should be used for calibration if a true molecular weight distribution is required.

If no primary standards are available two approaches can be taken, either polystyrene standards are used to provide an apparent molecular weight or alternatively for homopolymers and copolymers the 'universal calibration' can be employed. This method developed by Benoit utilises the fact that the elution volume of different polymer types in the same solvent is proportional to their molecular weight and their intrinsic viscosity. The intrinsic viscosity can be obtained either by a viscometer detector or be calculated from the Mark-Houwink equation

$$[\eta] = KM^{\alpha} \qquad (9.6)$$

where K and α values for many polymer/solvent systems are available in the literature. Thus, providing information is attainable for $[\eta]$ of the polymer and polymer standards, absolute molecular weight can be determined.

9.3.5.2 Preparative SEC. Size exclusion chromatography is primarily directed at detecting the concentration of each different molecular weight or mass present in the polymer. This approach is quite straightforward for linear homopolymers, however if the polymer is a linear copolymer then the molecules can vary in composition and sequence length as well as molecular weight.

The analysis of such systems is often complex. One solution to this problem is the use of preparative SEC, to provide fractions for subsequent analysis by off-line techniques such as mass spectrometry, Fourier transform infrared and nuclear magnetic resonance. This technique can also be used to isolate pure polymer fractions for subsequent use in SEC calibration.

Normal analytical scale SEC equipment can be used, with the essential additions of a high volume injector and to reduce operator time, an automatic fraction collector. The columns used contain typical SEC packing materials but are normally 25 mm internal diameter and 30 or 60 cm in length. Such columns offer up to a tenfold increase in loading over normal 7 mm analytical columns. Flow rates of $10 \, \text{ml min}^{-1}$ of eluent are normally used. Fractions are taken at specified points in the molecular size distribution, as monitored on the RI detector at the end of the column. The eluent from the detector is then collected by the fraction collector into designated sample tubes. These fractions are then taken to dryness and examined using other techniques. Since an SEC separation is normally achieved irrespective of the solvent used, volatile solvents can be chosen for SEC analysis to aid fraction isolation. Larger columns (57 mm × 122 cm) are commercially available, but these require more specialised and dedicated preparative instrumentation as flow rates in excess of $100 \, \text{ml min}^{-1}$ are required for fractions. These systems can offer higher sample loading on the columns.

9.3.5.3 Separation of low molecular weight species. The molecular weight/ mass distribution is not the sole factor affecting the physical and processing characteristics of a polymer. The presence of residual low molecular weight species such as monomers, starting materials and oligomers can have an affect, and in the case of formulated systems the additives such as plasticisers, heat and light stabilisers, antioxidants and lubricants are by nature added to promote or suppress effects in a polymer system.

Separation of such low molecular weight components can in many cases be accomplished by non-SEC technology (section 9.4). However SEC using highly efficient 3 μm and 5 μm columns can often offer an easier alternative.

'HPLC' separation of polymers requires the removal of polymer prior to the analysis, low molecular weight SEC can operate without such need, as the polymer will just be excluded from the system. SEC presents a quick and easy method development strategy; the process involves the selection of a suitable solvent and the use of any small pore size columns 50 Å and 100 Å. All components will elute within a predetermined interval, amounting to one column volume, and the elution volume of any material can be predicted for a given column provided a calibration curve is available.

SEC for low molecular weight separations is ideally suited for quality control, quantitation, separation of unknown and removal of polymeric species from a system.

9.3.5.4 Aqueous SEC. A number of synthetic polymers require the use of aqueous based mobile phase systems for separation/characterisation. These include polyacrylamides, polyvinyl alcohols, poly(sodium vinyl

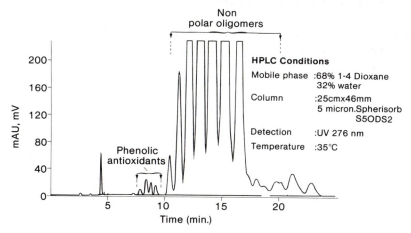

Figure 9.6 HPLC analysis of phenolic antioxidants in low molecular weight acrylic resin.

sulphonates), polyacrylic acids, polyethylene glycols, hydroxyethyl cellulose, polymethacrylate salts and polyethyleneimine.

As with non-aqueous SEC, gel-based column packings are commonly used (based on polyacrylamide, polyhydroxyl or polymethacrylate). They are capable of operation in:

(i)   water
(ii)   water plus salts
(iii)   water plus organic modifiers
(iv)   buffers, within a pH range 2–10
(v)   temperatures up to 80°C

The column compatibility with a wide range of aqueous-based solvent systems is essential as in aqueous SEC it is often necessary to suppress sample/column interactions to preserve a pure SEC mechanism.

The choice of eluent system depends on the polymer type. For most non-ionic hydrophilic polymers, water can be used. However much more complex eluent systems are needed, for anionic and cationic polymers where interactions with the column based on ion exclusion, inclusion and exchange, adsorption by hydrogen bonding or hydrophobic interactions and intramolecular electrostatic effects, are possible. This can often make method development in aqueous SEC extremely difficult and time-consuming.

Further problems arise with the technique by the lack of availability of suitable primary calibration standards and the high viscosity of the synthetic polymers in water, resulting in much lower sample concentrations being used for injection (as low as 0.01%).

9.4 Other modes of chromatography

As stated in the Introduction, the most widely used LC separation mode for polymer analysis is SEC. For low molecular weight species, however, where separation of similar size molecules is required, interactive liquid chromatographic modes (partition/absorption) are more suited.

These non-SEC modes use high efficiency silica-based columns and binary, ternary and quaternary miscible mixtures of organic solvents and water to achieve fast, selective separations (Figure 9.6). In general, a reverse phase mode of operation is most suitable for lipophilic samples, whilst normal partition or adsorption is used for samples which are lipophobic.

The main use of these more 'typical' HPLC modes is in the analysis/ quantitation of polymer additives:

— plasticisers (phthalates and adipates)
— antioxidants (phenols and amines)
— surfactants (non-ionic and anionic)
— stabilisers (alkyl, hydrox phenols)

Catalysts (peroxides) and monomers/oligomers (acrylate and epoxy) can also be separated.

For the reasons described in section 9.2, it is advisable to remove all polymer prior to any such analysis. The equipment used for such determinations can be described as conventional HPLC hardware; the main difference from SEC being that solvent pumping system are normally employed that can generate solvent gradients with time and more sophisticated UV detectors (diode array) are often utilised which can provide not only quantitative data, but identification/peak purity information on the separated peaks.

Bibliography

Benoit, H., Grubisic, Z. and Rempp, P. (1967) *J. Polym. Sci. B* **5**, 753–759.

Evans, J.M. (1973) *Polym. Eng. Sci.* **13**, 401–408.

Giddings, J.C. (1967) *Anal. Chem.* **39**, 1027–1028.

Glöckner, G. (1987) In *Polymer Characterisation by Liquid Chromatography, Journal of Chromatography Library*, Vol. 34, Elsevier, Amsterdam, pp. 404–405.

Janca, J. (1984) In *Steric Exclusion Liquid Chromatography of Polymers*, Marcel Dekker, New York, pp. 1–47.

Jordan, R.C. (1980) *J. Liquid Chromatogr.* **3**, 439–463.

Kuo, C.Y., Provder, T., Koehler, M.E. and Kahn, A.F. (1987) In *Detection and Data Analysis in Size Exclusion Chromatography*, ed. T. Provder, ACS Symposium Series No. 352, American Chemical Society, Washington, DC, pp. 130–154.

Moore, J.C. (1964) *J. Polym. Sci. A (Gen. Papers)* 835–843.

Porath, J. and Flodin, P. (1959) *Nature* **183**, 1657–1659.

Vaughan, M.F. (1960) *Nature* **188**, 55.

10 HPLC in biomedical and forensic analysis

D. PERRETT and P. WHITE

10.1 Introduction

Biomedical and forensic analysis are dealt with together in this chapter since for both of these application areas, the chromatographer encounters very similar types of samples. Biomedical analysts work with samples of biological fluids which is also the type of sample with which the forensic analyst primarily works. Other types of forensic samples are also often of a similar nature, in that low levels of analyte in a complex matrix are involved. However, in looking at biomedical and forensic analysis together, it will be observed that by virtue of the differences in analytical problems posed, the nature of the HPLC carried out may often be quite different.

10.2 Biomedical analysis

The analysis of chemical components in clinical samples is primarily used to diagnose disease states and to monitor the progress of the therapy of a disease. It may also be used to further our understanding of normal and disease processes in the human body through biomedical and therapeutic research. There are many techniques available to the clinical chemist to achieve these measurements and it is probably fair to say that although many clinical chemistry laboratories may possess HPLC equipment, an equal number do not and in many laboratories where it is available, it is under-utilised. This is because clinical chemistry laboratories are used to dealing with many hundreds of simple assays per working day; the equipment to perform these assays is in constant use and rapid turn round times are essential. This means that chemistries must be very robust. Sample preparation should be minimal and if possible assays should work with neat plasma or urine. For techniques such as enzyme-linked immunosorbent analysis (ELISA) assays, many tens of samples can be performed in parallel. Assays should be designed such that staff can perform them with relatively little specific expertise. It may also be mentioned that in many cases, the precision of the individual measurement can be relatively low since normal variation can be large and it is unlikely that threshold values need to be analytically precise. Data output as a

Table 10.1   Principal areas of application of HPLC in clinical chemistry

Area	Example(s)
Therapeutic drug monitoring	Theophylline cyclosporin levels following transplants
Metabolite assays	Vitamins, neurotransmitters, e.g. catecholamines
Metabolic profiling	Amino acids in genetic disease purine and pyrimidines in genetic disease
Protein assays	Haemoglobin A_{1c}
Enzyme assays	Cytidine deaminase xanthine oxidase
Purifications	Labels for immunoassay
Definite assays	

single number is preferred for ease of interpretation by both the clinical chemist and the clinician.

HPLC does not necessarily fit these criteria. It is a serial procedure, one sample being injected when another has finished and is typically capable of only 6–10 assays per hour. Sample preparation for HPLC is often complex and time-consuming. A moderate degree of expertise is essential to operate an HPLC system satisfactorily and to interpret the chromatogram that is produced. The chemistry of the system usually changes slowly but is sometimes capable of quite rapid change. The results produced may occasionally even disagree with more traditional and established assays leading to a conflict of data and clinical meaning. HPLC is relatively expensive both to acquire and to operate.

Nevertheless HPLC has found a place in many routine clinical chemistry laboratories and is almost essential in a biomedical research laboratory today. It is very adaptable and can be used with a wide range of compounds, particularly those that are water soluble such as most metabolites. It can be automated to allow moderate throughputs e.g. about 100 samples per day. With appropriate detectors, assays can be very sensitive; very accurate assays can be developed and quantitative precision can be very good. HPLC can also be preparative enabling physical identification of peaks. Table 10.1 lists some of those areas where it can be and is used in clinical chemistry.

Although therapeutic drug monitoring may seem an ideal application for HPLC it is not uncommon to find that an HPLC assay is developed in the first instance since this can be done relatively rapidly but if the assay proves clinically useful then an immunoassay which will take much longer to develop will be perfected. This same scenario has also been observed with other assays such as specific proteins since immunoassays are considered preferable when high throughput is necessary. When more than one compound is required, such as a drug and its metabolites or all of the 20 plus amino acids in plasma, or the three principal catecholamines,

HPLC is at its best. One HPLC run can measure all 20 amino acids but 20 different immunoassays would be necessary.

10.2.1 Method development for biomedical HPLC

Whatever the analytical problem being tackled (not just in the forensic and biomedical fields), it is essential to keep an overview of the aims in mind. The principle goals for the solution to any analytical problem should be:

- appropriate sensitivity
- maximum speed, i.e. minimum analytical time or maximum through-put
- maximum selectivity
- appropriate yield (if preparative or further analysis required)

Unfortunately it is rarely the case that all these goals can be maximised for any particular analysis. In analytical HPLC some may even be mutually exclusive, e.g. maximum resolution requires more time and may reduce sensitivity since gradient elution will be required. It is important to remember that the column is only one part of the overall system. For example, maximum selectivity can be achieved not only by attaining good resolution but also by choice of extraction conditions and correct choice of detector. Fluorescence and electrochemical detectors are not only considerably more sensitive towards appropriate analytes but also more selective than UV or RI detectors for many compounds. Higher detector specificity may simplify sample preparation and therefore improve throughput. Nevertheless the aim of any good analytical HPLC method is to completely resolve the compound(s) of interest with a reproducible chromatographic system in the least possible time and quantify the peak with adequate sensitivity, accuracy and precision.

A word of caution is necessary. Given the success of HPLC, it is easy to think that HPLC is always the best and most appropriate method of analysing most compounds in most samples. The good analyst must choose the best solution to each analytical problem basing the choice on the nature of the analyte, the level of accuracy and precision required, the expected numbers of samples as well as the facilities and expertise available. There is little purpose in spending months to develop a very precise assay when the sample collection procedure is imprecise or using HPLC to measure compounds more appropriately measured by an immuno-assay. Even given that a problem requires a separation mode, HPLC is not always the most appropriate method; for example a number of samples may be run in parallel by TLC if only qualitative information is required.

10.2.2 HPLC method development in biomedical analysis

Given a sample to analyse for a known or suggested compound or group of compounds, how is a HPLC procedure, whether analytical, preparative or a combination, established? The following lists suggest some of the preliminary observations the analyst should make.

1. What sort of answers to the problem are required?
    - Is the method to be analytical or preparative?
    - If analytical, is it quantitative or qualitative?
    - High throughput or maximum detail?
    - Can the chosen method be destructive?
    - To what level must the assay be validated?
    - Has the analyte been separated before; is a method in the literature?
2. Study your compound(s), its chemistry and its occurrence.
    - What is the molecular weight?
    - Does it isomerise or exist already in different isomeric forms, and is that composition required?
    - In what is it soluble?
    - Will the sample be limited in amount?
    - How will it be detected?
    - What column/eluent will be appropriate?
    - What sort of isolation and/or sample preparation procedure may be necessary?

Answers to all or most of these questions are necessary before one can develop any assay.

10.2.3 The organisation of HPLC in a biomedical laboratory

The aim must be to operate robust assays using the simplest possible configuration of HPLC instrumentation. Robust systems will only result from the proper development of an HPLC separation. Method development must involve not just pure standards but also a pooled sample. Much time can be wasted if the presence of an endogenous interference is not recognised at an early stage in method development. Time must be allowed for an investigation and full documentation of the effect of chromatographic variables such as pH, % organic modifier on the separation, selection of appropriate sample preparation procedures and then a full validation of the assay. Efforts should be made to reduce baseline noise and eluent/sample artefacts in order to maximise signal-to-noise responses. With this information on record it is easier to select stable conditions and, importantly, the details are available for easy trouble-shooting in the event of later problems.

If at all possible work should be carried out with isocratic elution since the equipment is simpler and cheaper, elution conditions more controllable and it is more economic since eluents can be recycled. Methods should be developed using the shortest columns that give the resolution required. Short (10 cm × 4.6 mm) columns packed with 5 μm bonded silica phases will resolve most analytes in most samples. For ease of operation and when high throughputs are required, auto-injectors and computer integration should be used although such sophistication should probably not be employed during method development. Manual injection is satisfactory when relatively few samples are to be run which is often the case with gradient elution assays. Gradient elution, although essential for some assays such as amino acid analysis, is more expensive in terms of equipment, operationally more complex, solvent costs are higher and elution times longer and therefore sample throughput lower.

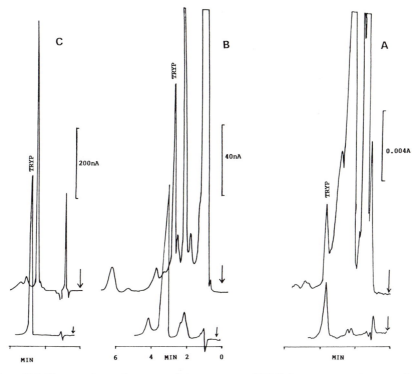

Figure 10.1 Comparative performance of three common HPLC detectors towards tryptophan in the same acid-extract of plasma (top trace) and a 400 pmol injection of tryptophan standard lower trace. (A) UV detector at 280 nm, (B) electrochemical detector at 0.8 V versus Ag/AgCl and (C) Fluorescence detector with deuterium lamp set at E_x 222 nm and $E_m > 330$ nm. The separation was performed on a simple RPLC system at pH 5.

Although the UV/vis detector is the most versatile detector used with HPLC it is not always ideal since its lack of specificity means higher resolution of the analytes may be required. It is also of only moderate sensitivity. Diode array and rapid scanning detectors are becoming increasingly popular, particularly as their price and the price and speed of the associated computer also fall. Such detectors are useful for peak identification and monitoring peak purity but they are somewhat less sensitive than single wavelength detectors and their additional cost is probably not justified for most routine applications. Both higher sensitivity and higher selectivity can be achieved with fluorescence and electrochemical detection of suitable analytes. If at all possible, fluorescence is to be preferred since such detectors are easy to operate, sensitive, stable and selective. Its selectivity shows itself in the lack of frontal components observed in plasma extracts whereas electrochemical detection is nearly always associated with a major frontal peak that tails considerably (Figure 10.1).

10.2.4 Samples and sample preparation for biomedical HPLC

The bioanalyst can be required to analyse most biofluids although the most common are urine and the aqueous phase of blood, i.e. plasma or serum. Other samples may be cell and tissue extracts, synovial fluid, cerebrospinal fluid (CSF) and saliva. In the case of urine and CSF with their very low protein content it might be possible to directly inject the sample into an HPLC column. With most silica-based packing materials, direct injection of blood proteins will rapidly lead to column deterioration. HPLC columns are expensive and their efficiency is easily lost so correct preparation of samples will not only improve column life but also improve the results. At its simplest it is only necessary to remove particulate matter from samples to prevent clogging of the column and frits. Modern HPLC packings are very susceptible to contamination by proteins, fats and other macromolecules from biological samples and it is necessary to remove these (except of course for protein analysis).

Extraction techniques should fulfil the following criteria to be ideal.

1. It should release all of the analyte(s) of interest from the sample including any that is protein bound.
2. It should precipitate all proteins in order to avoid detrimental chromatographic effects.
3. It should neither degrade the analytes chemically nor permit biochemical degradation. It should not degrade polymeric species and thereby increase the concentration of monomers.
4. The extractant should not interfere with the assay or be easily removed prior to analysis.
5. It is advantageous to selectively extract the compounds of interest

from others with similar chromatographic and/or detection characteristics.

6. The final extract should be clean and free of particulate matter.
7. The procedure should not dilute the extract more than necessary.
8. The compound(s) should be stable in the final extract, particularly if
    they are to be stored in it prior to analysis.
9. The total procedure should be simple, rapid, reproducible and give
    good recoveries.

For protein free samples such as urine, it is often only necessary to
either centrifuge the sample at high speed or pass it through one of the
0.45 μm filters especially designed for HPLC work. The particulate-free
supernatant may then be injected directly onto the HPLC column. The
preparation of plasma and serum samples can be much more complex. A
large number of techniques for the removal of proteins and fats exist and
some have been adapted for HPLC. The commonly employed methods
and their mode of action are shown in Table 10.2.

A useful study of the efficiency of some of these procedures for the
removal of protein from human plasma is presented by Blanchard (1981).
However, it is always necessary to check standard methods for your own
particular samples.

10.2.4.1 Solid phase extraction. With the availability of pre-prepared
cartridges of silica-based adsorbents, the use of solid phase extraction has
increased in the last few years although the technique has been in use for
many years for the isolation of many biochemicals, e.g. amino acids,
catecholamines. In essence it is a version of chromatography; conditions
for the selective adsorption of the analytes (column, solvent, pH, etc.) are
chosen, the sample is applied to a column, washed and the analytes selectively eluted with appropriate solvents. Since the columns are disposable
there is no need to worry about protein contamination or infection. The
adsorbents available cover an even wider range than HPLC materials
since they are not required to withstand high back pressures. It is possible

Table 10.2 Deproteinating procedures and their mechanisms of action

Procedure	Principle
Heating	Denatures proteins
Cold	Denatures membranes
Ultrafiltration	Molecular size separations
Ammonium sulphate	Salting out of analyte
$BaSO_4/Zn(OH)_2$	Co-precipitation
Solid phase extraction	Selective adsorption and desorption
Mineral acids	pH denaturing
Methanol and other solvents	Dielectric changes solubility

to use chemical reactive adsorbents such as alumina for diol groups. When possible, it is always preferable to employ a phase that is complimentary to the analytical system, i.e. cation exchange clean-up followed by analytical RPLC. Unlike analytical columns, a considerably greater amount of sample can be loaded onto clean-up columns.

10.2.4.2 Use of internal standards. With complex extraction procedures, it is preferable to use an internal standard to correct for both recovery and analytical variability. The internal standard should be chemically similar to the compounds of interest and be chemically stable. It should have similar detection characteristics as the unknown and should elute in a blank portion of the chromatogram preferably near the middle (or near specific peaks of interest), and be well resolved from adjacent peaks. To accurately reflect the recovery of the unknowns it must be added as near as possible to the start of the extraction procedure. The internal standard corrects mainly for dilution and sampling errors. The amount of internal standard added should be such that allowing for dilution it will give a prominent well resolved peak with a height of about 80% full scale under normal analytical conditions.

10.2.5 HPLC of ionic compounds

Unfortunately for the bioanalyst, many naturally occurring molecules such as amino acids are highly charged and in general such molecules are not well retained by the hydrophobic reversed-phase materials most popular in HPLC. Ionic compounds can be separated by ion-exchange HPLC, but silica-based packings have relatively low efficiency and exhibit a tendency to lose ion-exchange capacity with use. They may also require considerable ionic strength and/or pH gradients. The more ionic the molecule the faster it elutes from an ODS column. Techniques which improve retention of ionic species in RPLC are ion suppression and ion pairing.

10.2.5.1 Ion suppression. By controlling the ionisation of weakly ionic samples they are retained on reversed phase columns; for example, a weak acid is retained using a eluent with a pH below the pK_a of the acid. But silica-based materials can only normally be used between pH 1.5–7.5 so increased retention for bases cannot usually be obtained in this way.

10.2.5.2 Ion pairing. Correct choice of eluent pH encourages ionisation of the sample which then pairs to an oppositely charged soap-like counter-ion dissolved in the eluent, which leads to increased chromatographic retention of the 'non-ionic' pair. The counter ion (e.g. octanesulphonic acid, tetrabutylammonium hydrogen sulphate) usually includes in its structure a strongly hydrophobic moiety. Other mechanisms can be equally if not more important, such as the binding of the ion-pair itself to the hydrophobic silica surface creating a dynamic ion-exchange surface. Whatever the mechanism, ion-paired RPLC is very important for the separation of many classes of biopharmaceutical compounds.

10.2.6 High sensitivity assays

Particularly when working with endogenous biomolecules the analyst will often be working near the limits of sensitivity of the instrumentation and it is therefore very important to optimise the chromatographic system if the assay is to be at all robust. Of course a most important aspect is the correct choice of high sensitivity detector. For high sensitivity assays, it is essential to use isocratic conditions with the peak(s) of interest eluting with a k' of 3–5 since it is only then that detectors can operate at their maximum sensitivity. The peaks should be well resolved from the frontal peak and other possible interferences. If at all possible the selectivity should be adjusted so that any small peaks to be measured elute before large peaks and do not sit on the tail of large peaks. Care must also be taken with regard to the choice of eluent and its purity and correct flow rate selection as well as good sample preparation for maximum signal-to-noise ratios to be obtained.

Even with all these chromatographic parameters optimised, sensitivity may not be enough. The simplest way to increase sensitivity further is to inject more sample. Valve injectors are usually supplied with 20-μl loops but there are few assays for which 100–200 μl of sample cannot be injected simply by changing the loop size. A further way is to use minibore (1 mm or 2 mm i.d.) or even microbore (<1 mm) columns. Such columns give increases in sensitivity due to the narrower peaks eluted but this may reduce column loading volumes and with some equipment the extra sensitivity gained may be lost due to extra column band broadening.

10.2.7 Detection of weakly absorbing compounds

Often it is required to detect compounds with no or only very weak chromophores such as sugars and amino acids. Refractive index detectors and mass sensitive detectors can be used but they are relatively insensitive in the context of biological sample concentrations. Indirect detection using a UV or fluorescent eluent can also be employed. However, the most common approach is the use of derivatisation. Derivatisation of some chemically reactive moiety on the analyte can be performed in two modes. In post-column derivatisation the sample is separated first and then reacted with a flowing stream of derivatising reagent being pumped into

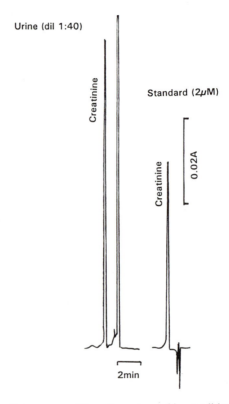

Figure 10.2 Urinary creatinine assay. The chromatographic conditions were as follows: column, 150 mm × 4.6 mm (5 μm) cation-exchange silica (SCX HPLC Technology, UK); eluent, 50 mM sodium formate adjusted to ca. pH 6–methanol (80:20); flow rate, 1 ml min^{-1}; injection, 20 μl; temperature, ambient; detector, variable wavelength UV detector (Cecil) set at 230 nm. Chromatogram courtesy Dr Ian James and the author, Department of Medicine, St Barts. London.

the eluate at the base of the column. In pre-column derivatisation the whole of the sample is first reacted with the reagent and then the reaction mixture is injected on to the HPLC column. Clearly such a procedure will change the nature of the analyte but this can often be advantageous since it will usually increase its hydrophobicity and improve its chromatography. The choice of derivatisation mode is dependent upon factors too numerous to deal with here.

10.2.8 *Typical applications of HPLC in biomedicine*

Many of the points described above are illustrated by the following brief descriptions of a number of typical practical examples.

10.2.8.1 Creatinine in urine.   The assay of creatinine in body fluids is one of the core assays in clinical chemistry since its level in blood and urine reflects the functional status of the kidney. There are many methods for its assay ranging from the simple colorimetric Jaffe reaction to dedicated creatinine analysers using discrete sampling technologies. For many metabolic assays, the so-called creatinine correction can be applied since creatinine excretion is considered to be constant throughout the day. The clinic therefore only needs to collect a random specimen of urine rather than a full 24 h specimen.

The most accurate measurements of creatinine involve some form of separation, so an on-line method is very useful. Figure 10.2 shows a simple cation exchange isocratic HPLC assay that can be easily set up in a metabolic laboratory. The urine is simply diluted prior to injection. Creatinine is detected by its UV absorption but the figure clearly shows that trace amounts of other compounds must be resolved if a precise assay is to be achieved.

10.2.8.2 Catecholamines.   The catecholamines, adrenaline, nor adrenaline and dopamine, are important neurochemicals that occur at trace levels in human plasma ($0.05–2\,\text{pmol}\,\text{ml}^{-1}$) and at about 100-fold those levels in urine. Although they are UV-absorbing at circulating levels in plasma, UV detection is grossly insensitive but being phenolic they are electrochemically active and can be detected by oxidation at a glassy carbon electrode maintained at approx. $+0.7\,\text{V}$ versus an Ag/AgCl reference electrode. The best electrochemical detectors are capable of sensitivities of below 100 fmol injected. Even so, there are many other electrochemically active compounds in plasma and urine and a selective extraction step is necessary to reduce the level of interferences and in the case of plasma also give some increase in concentration. Catecholamines can be selectively adsorbed onto alumina from plasma and urine and then desorbed by acid prior to injection. To improve precision and accuracy an internal standard is necessary. Finally catecholamines are cations and only poorly retained on RP columns so their chromatography requires the use of either an ion-pairing agent such as octanesulphonic acid or less commonly a cation exchange separation. A typical plasma separation is shown in Figure 10.3.

10.2.8.3 Indoleamines in urine.   5-Hydroxyindolacetic acid (5-HIAA) is the principle metabolite of the neurotransmitter 5-HT and is excreted in urine. It is naturally fluorescent (excitation 280 nm and emission 360 nm, using a xenon lamp). The selectivity of fluorescence detection means that a very simple and rapid assay is possible. Provided the pH of the eluent is controlled in the region 5.1–5.4 (the exact value depends upon the make of C18 columns being used) 5-HIAA can be resolved from the front and

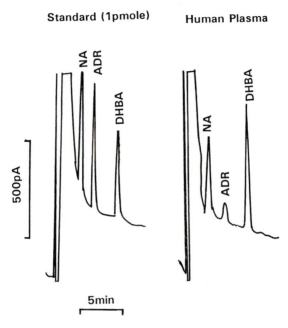

Figure 10.3 Separation of plasma catecholamines using ion-paired reversed-phase HPLC and with detected at high sensitivity using electrochemical detection. The chromatographic conditions were: column, 100 mm × 4.6 mm (5 μm) ODS Hypersil; eluent, [40 mM potassium dihydrogen phosphate, 40 mM; trisodium citrate, pH 6.4) containing 5 mM octane sulphonic acid and 1 mM EDTA]–methanol (90:10; v/v); flow rate, 1 ml min^{-1}; injection, 50 μl; temperature, ambient; detector, electrochemical set at 0.7 V versus Ag/AgCl. Reference: Bouloux, P., Perrett, D. and Besser, G.M. (1985) Methodological considerations in the determination of plasma catecholamines by high performance liquid chromatography. *Ann. Clin. Biochem.* **22**, 194–204.

from tryptophan, a major fluorophore in urine. A separation is shown in Figure 10.4.

10.2.8.4 Nucleotides, nucleosides and bases in blood and urine. These related groups of compounds are important to normal cellular function but also exhibit abnormalities in diseases such as gout. There are some 20 common nucleotides of which adenosine triphosphate (ATP), adenosine diphosphate (ADP) and adenosine monophosphate (AMP) are the most important. Nucleotides consist of a hydrophobic purine or pyrimidine nucleoside to which is attached up to three phosphate groups. They occur within cells and must be extracted by acid precipitants before analysis. Being strong anions they can be separated by either anion exchanger or ion-pair HPLC but the number of individual nucleotides found in extracts and their diverse nature means that gradient elution is almost

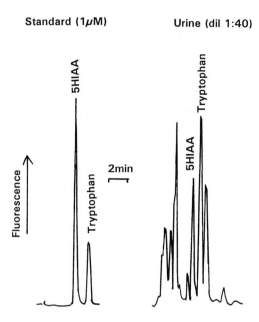

Figure 10.4 Assay of indoleamines such as 5-HIAA and tryptophan in urine by HPLC with natural fluorescence detection. The chromatographic conditions were: column, 100 mm × 4.6 mm (5 μm) ODS Hypersil; eluent, 50 mM ammonium acetate (pH 5.25)–methanol (250:30, v/v); flow rate, 1 ml min^{-1}; injection, 20 μl; temperature, ambient; detector, fluorometric (Jasco with xenon lamp E_x 280 nm E_m 335 nm).

essential. They absorb strongly in the range 250–260 nm. Figure 10.5 shows the gradient anion exchange separation of nucleotides in human red blood cells.

In contrast nucleosides are bases of moderate hydrophobicity. They are easily resolved by reversed phase HPLC but again their complexity in biofluids means gradient elution is usually required.

10.2.8.5 Amino acid analysis. There are some 20 amino acids found in proteins and these are released by overnight hydrolysis in 6 M HCl. Plasma and urine contain an even larger number of amino acids or related compounds. At low pH, amino acids are cations and for 40 years have been separated by cation exchange column, chromatography. The problem with amino acids is that in general they possess no chromophores by which they can all be detected. In the traditional amino acid analyser, their detection was accomplished by a post-column reaction with ninhydrin which forms a purple colour on heating with an amino acid at pH 5.5. This colour, Ruhemann's purple, is formed with all primary amino acids and can be detected at 570 nm. Secondary amino acids such as proline form a yellow chromophore measurable at 440 nm.

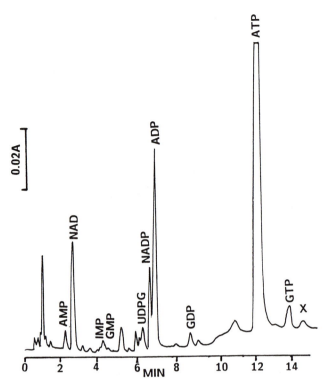

Figure 10.5 Separation of nucleosides in human red cells by anion exchange HPLC. Chromatographic details: column 4.6 × 100 mm (5 μm) APS-Hypersil (Shandon); detector, UV (Cecil CE2012) 254 nm; eluent, linear gradient generated on low pressure side of pump. 0.04 M KH$_2$PO$_4$; pH 3.0–0.8 M KH$_2$PO$_4$ (pH 3.2); flow, 1 ml min^{-1}; pressure, 400–500 psi; temperature, ambient; sample, trichloroacetic acid extract of human red cells (25 μl injected).

The arduous conditions for this reaction has meant that a number of pre-column derivatisation reagents for amino acids are now employed. Derivatisation of the amino group is readily achieved to form either chromophores, electrophores or most commonly, fluorophores. A popular chromophoric reagent is phenyl isothiocyanate (PITC) and both o-phthalaldehyde/thiol (OPA) and 9-fluorenyl methoxy carbonylchloroformate (FMOC) are commonly employed fluorescence reagents. There are many other reagents in the literature. All these reagents form unique products with each amino acid and the reaction products exhibit considerable variations in hydrophobicity, so it is necessary to use complex gradient elution schemes to resolve them all in one run. The pre-column fluorescence reagents are about 1000-fold more sensitive than the traditional amino acid analyser and about 100-fold more sensitive than the PITC reagent. However, not all reagents react equally with all amino acids, e.g.

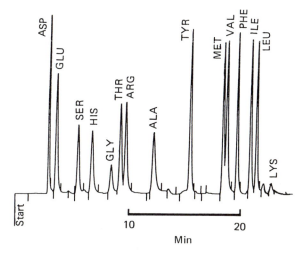

Figure 10.6 HPLC of OPA derivatives of amino acids. Chromatographic details: column, 4.6 × 150 mm (5 µm) ODS-Hypersil (Shandon); detector, Fluorescence (Kratos FS 970 E_x 230 nm E_m 360 nm); eluent, linear gradient from 50 mM phosphate (pH 5.5) to 100% acetonitrile; flow, 1 ml min^{-1}; temperature, ambient; sample: amino acid standard mixture derivatised with o-phthalaldehyde–3-mercaptopropionic acid injection equivalent to 59 pmol of each amino acid (25 µl loop).

OPA does not normally react with secondary amino acids. Figure 10.6 shows the separation of the common protein amino acids using OPA/3-mercaptopropionic acid reagent.

10.2.9 Progress in biomedical HPLC

Great strides have been made in biomedical HPLC, particularly in the last decade and it could now be considered a very mature technique. However, the technology of HPLC continues to change quite rapidly and some of these advances could be very relevant to biomedical separations. New packing materials such as polymerics, base deactivated silicas, pyrolysed carbon and internal surface packings should offer improved stability and higher efficiencies for certain classes of compound such as basic drugs. Microbore columns should become more accepted since they offer not only improved sensitivity but also lower solvent consumption and consequently reduced needs to dispose of noxious solvents. Chiral HPLC may become important for determining trace levels of racemic drugs in man. There are many other improvements in HPLC that can be discussed but the bioanalyst must remember that other analytical techniques such as immunoassays are also improving and even old separation techniques such as electrophoresis can, with new insights, quickly develop into modern high efficiency analytical tools such as capillary electrophoresis.

10.3 Forensic analysis

The potential for using HPLC in forensic science laboratories was recognised when the technique was in its infancy. This interest arose because of the difficulties encountered with the analysis of basic drugs, and it was soon to be appreciated that HPLC offered certain advantages over gas chromatography (GC). Once it was established that reproducible qualitative and quantitative analysis could be performed in several minutes there was a keenness to determine if HPLC could be used to solve other analytical problems experienced by the forensic scientist.

Within a forensic science laboratory, the analytical problems are often very different to those found in other laboratories. Extremely small and often aged samples, complex matrices and an extensive range of analytes are encountered by the forensic scientist and the success of any HPLC method is often very dependent upon the selectivity and/or sensitivity of the system. General developments in column and detector technology have played a major role in improving both these criteria and hence increasing the number of forensic applications of HPLC.

Selectivity, the ability to isolate a particular analyte or separate a number of components within a mixture, has improved dramatically through the development of bonded-phase column packing materials for reversed-phase, ion-exchange and ion-pair chromatography. More recently, forensic laboratories have been introducing polymeric packing materials because of their selectivity and other physical and chemical properties which are more desirable than those of silica-based materials.

In its broadest sense, selectivity has also been achieved through the development of ion chromatography and indirect photometric detection. Apart from opening up a completely new application of HPLC these provide highly selective techniques for the analysis of inorganic and organic ions.

Other methods for improving selectivity include the use of specific detectors and detection techniques, either with or without sample derivatisation. Therefore, electrochemical, fluorescence and multiwavelength UV/visible detectors feature strongly in forensic applications. Combinations of two or more different types of detector linked in series are also being used more frequently as a method of improving solute discrimination/identification. Finally, gradient elution techniques can increase solute resolution but this is often at the expense of longer analysis times which is a major consideration if large numbers of samples are to be analysed. Interestingly, despite the complexity of samples many forensic methods are performed very successfully under isocratic elution conditions.

Sensitivity, which is defined as a measure of the minimum amount of sample that can be detected, is often a major concern. Sometimes these problems can be resolved by employing a fluorescence or electrochemical detector and/or preparing a derivative of the analyte. True microbore (i.e. column i.d. < 1.0 mm) HPLC systems can provide gains in sensitivity but these are not used extensively in forensic laboratories mainly because of practical problems associated with reproducibility and short column life-times. However, there is a trend towards using 'narrow-bore' columns of 2–3 mm i.d.

An important point that must not be overlooked is that enhanced sensitivity and selectivity can often be achieved by ensuring a suitable sample extraction procedure has been adopted. Ideally, extraction should yield a high recovery of the analyte and a low level of co-extractives. Whilst every effort is made to achieve this in forensic laboratories, some toxicological samples, especially extracts from whole blood, liver and stomach contents, do present some difficulties. Unfortunately, the forensic toxicologist does not have the luxuries of working in a clinical laboratory where the subject is alive, larger volumes of body fluids can be obtained and plasma and urine can probably be used for analyses. However, an important point to consider, especially if working in an operational laboratory, is that cleaner extracts will generally improve the lifetime of an HPLC column.

The success of HPLC in forensic science is not due solely to these general developments. In-house research and development work has made a significant contribution and this has been, and still is, essential because of specific problems which are experienced in the analysis of casework samples. As a result of these developments, HPLC is now used extensively for the analysis of drugs, metabolites, rodenticides, anions, sugars, dyes, polymers, optical brighteners, explosives, fatty acids and other miscellaneous compounds.

Some details are now provided to show how HPLC is used for the analyses of these various classes of compounds. This is not intended to be a major review and topics have been selected to illustrate the range of techniques employed, and in particular how selectivity and sensitivity problems can be resolved.

10.3.1 Analysis of drugs of abuse

In most forensic laboratories, the largest amount of HPLC time is taken up with the analyses of drugs of abuse (often called 'street' drugs) i.e. amphetamines, heroin, cocaine, LSD, etc. The purpose of any drug's HPLC analysis is to confirm the identity of drug and provide a quantitative result. With the exception of LSD, relatively large quantities of a drug sample are usually available and UV detection is preferred since it offers

LSD   MORPHINE   DIAMORPHINE

COCAINE $\triangle^9$-THC   AMPHETAMINE

more than adequate sensitivity. Due to the excellent linearity of this type of detector, rapid quantitative analyses can be performed against a single standard of the drug being determined. However, it is essential that some form of quality assurance should always be included when this type of analytical procedure is employed.

As mentioned previously, the analysis of amphetamines and other basic drugs was one of the first forensic applications of HPLC. Many analyses of basic drugs are still performed by the same method of ion-exchange chromatography on unmodified silica columns with an eluent buffered to about pH 9. Dissolution of the silica in the analytical column is prevented by using a pre-saturation column which is packed with a 40 μm particle size silica. This type of column is situated in the eluent line between the pump and the injection valve and should always be used when performing analyses with aqueous eluents on silica-based columns. By using these columns the lifetime of the analytical column can be extended considerably.

The choice of buffering reagents used in any system is important and to gain the best sensitivity and selectivity for basic drugs, ammonia/concentrated hydrochloric acid is favoured. By using this buffering system, the UV detector can be operated at wavelengths down to 220 nm and hence amphetamines, methadone, cyclizine, dipipanone, quinine and other basic drugs can be analysed both qualitatively and quantitatively.

The analysis of heroin samples gives rise to a different set of problems. Heroin contains a mixture of opiate alkaloids and one of these, diamorphine (diacetylmorphine), is the controlled drug which has to be determined quantitatively. The other major opiates present in heroin can include codeine, acetyl codeine, mono-acetyl morphine and morphine and these have to be separated from diamorphine. Initially this presented some difficulties because of their similar chemical structures, but separation of these opiate alkaloids can be obtained either by ion-exchange chromatography on unmodified silica or ion-pair chromatography on reversed-phase packing materials. UV detection provides adequate sensitivity and by monitoring at 275–280 mm greater selectivity can be achieved for the opiates.

LSD is an example of a drug where dosage is extremely low and a detection limit of less than 100 ng must be obtainable. This excludes UV as a detection method, but fortunately LSD has a native fluorescence and therefore this detection level can be obtained very readily with a fluorescence detector. Again it is important that LSD can be resolved from other ergot alkaloids and this can be achieved on unmodified silica by ion-exchange chromatography.

For a very high proportion of cases, samples tend to contain basic drugs but some acidic and neutral drugs are encountered. Neutral or weakly acidic drugs (e.g. barbiturates) can be chromatographed on a reversed-phase system whilst acidic drugs (e.g. paracetamol, cannabis, etc.), are separated either by ion-suppression or ion-pair chromatography on a reversed-phase packing material.

UV is once again the preferred detection method for these drugs and monitoring wavelengths are selected to achieve the best sensitivity and selectivity for the drug being analysed. In the case of cannabis samples, dual wavelength monitoring at 220 and 254 nm is often used to identify and quantify the controlled drug Δ^9-tetrahydrocannabinol (Δ^9-THC). This procedure is favoured because cannabis samples contain many cannabiniods and greater selectivity can be achieved by employing this technique. Dual wavelength monitoring can be performed by either using two UV detectors coupled together in series or a multi-wavelength detector.

Since multi-wavelength detection has now been mentioned and will feature quite prominently in other applications, it is worthwhile discussing the reasons for this and why forensic laboratories have been keen to use and develop new multiwavelength detection techniques. Traditionally, solute characterisation in HPLC is based upon the measurement of retention time. However, for the purpose of sample identification or discrimination retention time is a very poor parameter because of the possibility of this being similar or identical to other compounds.

Multi-wavelength detectors permit UV or UV-visible wavelength ranges to be scanned rapidly and therefore simultaneous plots at several wavelengths can be obtained and spectra of the analyte can be generated at very short time intervals during an entire chromatographic run. These detectors require a microcomputer for their operation and this allows all

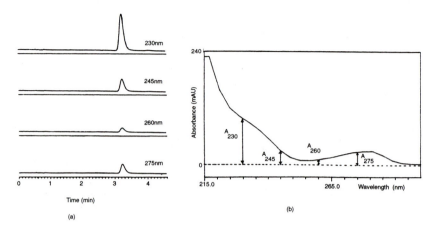

Figure 10.7 (a) Multiwavelength plot of a drug sample and (b) UV/visible spectrum of the analyte eluted with a retention time of 3.3 min.

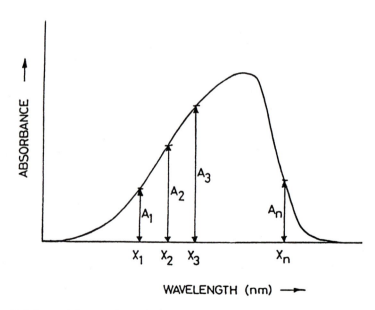

Figure 10.8 Spectral data used to calculate a PPP value within a wavelength range of x_1 to x_n nm. See text for an explanation of the symbols.

the chromatographic and spectral data to be stored. A number of routines have now been developed to evaluate these data and hence dramatic improvements in the characterisation of solutes can now be achieved. An additional advantage, and an important consideration if only a limited amount of sample is available, is that this information can be achieved from a single chromatographic analysis.

The analyst can improve the identification of solutes or discrimination between samples by studying the multi-wavelength plots or spectra of the eluted components. Transformation of spectral data to yield derivative spectra have been used to improve sample discrimination, whilst a plot of the ratio of absorbances for two selected wavelengths is an excellent method for confirming the homogeneity of chromatographic peaks. However, there is a trend in forensic laboratories to use and develop techniques whereby the spectral data are reduced to a smaller number of numerical values. Results in this format at easier to compare and present in court and furthermore can be evaluated statistically.

Absorbance ratioing and peak purity parameter (PPP) are examples of these numerical routines used for the analysis of drugs and other UV absorbing species. The principle of the absorbance ratioing technique using either the chromatographic or spectral data generated with a multi-wavelength detector is illustrated in Figure 10.7.

By selecting one wavelength as the reference wavelength, the absorbance ratios can be calculated from the chromatographic data used to construct the multiwavelength plot (Figure 10.7a) *viz.*, if 275 nm is the reference wavelength selected the absorbance ratios will be A_{275}/A_{260}, A_{275}/A_{250} and A_{275}/A_{240}. Similarly, if the spectrum of a solute is obtained as shown in Figure 10.7b absorbance ratios can be calculated from the spectral data.

The peak purity parameter (PPP) was introduced by an instrument manufacturer (Varian) and the PPP value is derived from the spectral data obtained over a wavelength range by using the following algorithm:

$$\text{PPP} = \frac{A^2_1 x_1 + A^2_2 x_2 + \cdots + A^2_n x_n}{\sum\limits_{x=1}^{x=n} A^2} \text{ nm} \qquad (10.1)$$

where x_1, x_2 and x_n are the wavelengths monitored by each of the diodes and A_1, A_2 and A_n are their respective absorbance values as depicted in Figure 10.8. PPP values can be obtained for one or any number of selected wavelength ranges.

An example illustrating the use of these numerical techniques namely, absorbance ratioing, is shown in Figure 10.9, where the controlled drug, diamorphine (peak X), is identified from retention time and absorbance ratio data at trace levels (~ 50 ng) in washings off a syringe and some silver foil.

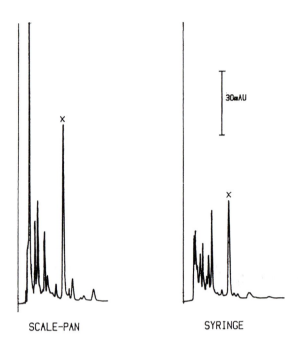

SAMPLE	RT	ABSORBANCE RATIOS		
	(min)	275:260	275:245	275:230
SCALE-PAN	2.11	2.20	0.77	0.25
SYRINGE	2.12	2.21	0.75	0.26
DIAMORPHINE	2.12	2.20	0.77	0.26

Figure 10.9 Identification of diamorphine (peak X) in extracts obtained from a scale pan and a syringe by the use of retention time and absorbance ratio data.

10.3.2 Analysis of toxicological samples

Toxicology is the study of poisons (toxins) and indeed many compounds can act as a poison. The forensic toxicologist is required to identify and quantitate these in virtually any matrix. A high percentage of analysis performed are for drugs and their metabolites in body fluids, e.g. blood, urine, liver, stomach contents, etc. However, food, drinks and many

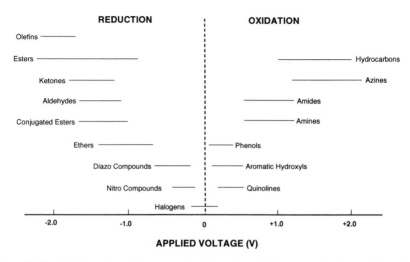

Figure 10.10 Applied voltage ranges required for the oxidation or reduction of functional groups.

varied sample types may also have to be analysed to confirm the presence of a poison.

To undertake the analyses of drugs and metabolites in body fluids the analyst is faced with several major problems. Firstly, due to the complex nature of the body fluid, the drugs must be isolated by an extraction technique which ideally should provide a relatively clean extract, and the separation system must be capable of resolving the drugs of interest from any co-extractives. Secondly, both selective and sensitive detection techniques are essential because of the small sample sizes available (typically less than 500 µl), and the extremely small quantities of drug often present.

In general, the analysis of acidic and neutral drugs does not present too many difficulties with detection because their therapeutic and, hence, overdose levels are relatively high ($\mu g\ ml^{-1}$ range). Therefore, many of these drugs can be monitored in an HPLC eluate with a UV detector.

On the other hand, detection difficulties are often experienced with basic drugs because their therapeutic levels are in the low ng ml^{-1} range. Some of the problems can be alleviated by employing either an electrochemical or fluorescence detector, since it is possible to achieve detection levels of sub-nanogram with these instruments, if operated under favourable conditions. Furthermore, both of these detection methods can enhance selectivity because analytes will only be detected if they contain an electrochemically active group or fluoresce.

To perform electrochemical detection the analyte should contain a functional group which is capable of either oxidation or reduction if subjected

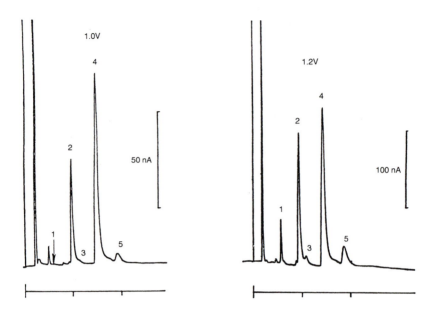

Figure 10.11 The effect of applied voltage on the response of dothiepin (2) and its meta-bolites, nordothiepin (1), nordotheipin *S*-oxide (3) and dothiepin *S*-oxide (5). Methdilazine (4) is used as the internal standard for this assay.

to a small positive or negative potential, respectively. Since functional groups will undergo either oxidation or reduction within a specific voltage range as shown in Figure 10.10, electrochemical detection is a highly selective detection technique.

Oxidative electrochemical detection has tended to be used more frequently, but this is only because it is more robust and easier to operate. Reductive analyses suffer from problems with oxygen interference and the requirement to use mercury as the electrode material.

Several HPLC oxidative electrochemical methods have been developed for the analyses of drugs and metabolites in body fluids and the quantitative analysis of morphine was one of the first forensic applications of this detection technique. By applying a potential of $+0.6\,\text{V}$ (versus an Ag/AgCl reference electrode) the presence of morphine can be confirmed in whole blood and a detection level of $10\,\text{ng}\,\text{ml}^{-1}$ for a 100-µl sample can be achieved routinely.

The development of a non-aqueous eluent system for ion-exchange separations on silica has provided an excellent system which, when used in conjunction with an electrochemical detector, permits the analyses of an extensive range of basic drugs and metabolites. This eluent system generates a very low background signal thus extending the operational range of the detector to about $+1.4\,V$ and hence a more extensive range of compounds can be detected. An example illustrating the use of this system for the analysis of dothiepin and its metabolites dothiepin S-oxide, nordothiepin and nordothiepin S-oxide is shown in Figure 10.11. The chromatograms illustrate the effect of the applied voltage on the solute response and how this parameter can be used to improve either detector selectivity or sensitivity.

Of the two detection techniques mentioned above, fluorescence is preferred in general because it suffers from fewer operational difficulties. Many compounds do not display a native fluorescence; however, this can be overcome if the analyte can be converted by chemical reaction into a fluorescent compound. This process is known as derivatisation and can be accomplished by either derivatising the sample prior to injection (pre-column) or after chromatographic separation (post-column).

A pre-column derivatisation technique using dansyl hydrazine is used for the quantitative analyses of glucose in blood or vitreous humour in cases of people suffering from hypoglycaemia and hyperglycaemia.

Post-column techniques tend to be more complex because additional instrumentation e.g. a pump and mixing chamber are required. However, some solutes can be made to fluoresce by simply modifying the eluant pH. This approach has been used for the analysis of quinine and rodenticides. With the former, the compound is chromatographed under basic conditions and then made fluorescent by the addition of perchloric acid after it elutes from the column. With rodenticides, chromatography under acidic conditions is needed and they are then made to fluoresce by changing the pH of the eluate to pH 11 by mixing with a methanol/ammonia solution.

10.3.3 Anions analysis

Organic and inorganic anions are encountered in forensic examination and qualitative and quantitative analyses are required. Since the introduction of ion chromatography, the analysis of several anions can be performed simultaneously and good sensitivity can be obtained with the conductivity detector.

An alternative method, indirect photometric detection (IPD) has also been used for anion analysis. Compared with ion chromatography, it can provide better discrimination between anions especially if the eluent contains citrate ions and is monitored at 220 mm. By using IPD under these conditions both positive and negative peaks can be generated as shown in Figure 10.13. For non-UV absorbing species, a negative peak will be detected (e.g. chloride, sulphate, phosphate, etc.) whereas UV absorbing anions will give a positive peak (e.g. nitrate, iodide, thiocyanate, etc.). Hence, anion identification is no longer based solely upon the retention of a peak because the peak direction provides additional information.

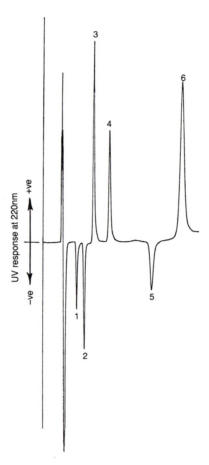

Figure 10.12 Indirect photometric analysis performed at 220 nm with an eluent containing citric acid. The positive peaks are due to the UV absorbing anions, nitrite (3), nitrate (4) and iodide (6), whereas the negative responses arise from the non-UV absorbing anions, phosphate (1), chloride (2) and sulphate (5).

Some anions can also be electrochemically active and an oxidative electrochemical detector is often coupled together in series with either the ion chromatograph conductivity detector or the UV detector when using the IPD technique. This dual detection technique can enhance anion identification or discrimination and provide increased sensitivity.

10.3.4 Dye analysis

Considerable evidence can be obtained from the analysis of colourants since dyes are encountered in a wide range of casework samples including

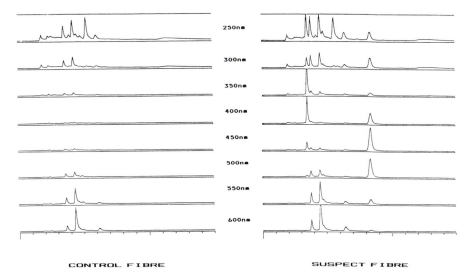

CONTROL FIBRE   SUSPECT FIBRE

Figure 10.13 Multi-wavelength monitoring of dye extracts from a single strand of a control and suspect fibre, both 5 mm in length. The chromatographic plots establish that the two samples could not have originated from the same source.

drinks, foods, plastics, fuels, drug powders and tablets, inks and fibres. These dyes can be either acidic, basic or neutral and silica-based column packing materials have been used to effect chromatographic separations. More recently, a polystyrene divinylbenzene (PSDVB) polymeric packing material has been used for the analysis of acidic dyes since it has been shown to provide better chromatographic properties and improve selectivity. An added advantage of using a PSDVB column is that there are no eluent pH restrictions.

Many of the HPLC dye applications require the analysis of trace levels of dyes. For example with the examination of fibre dyes, the casework sample is typically a single strand of fibre which is only 2–5 mm in length and the dye content is approximately 10 ng. Fortunately, because of the relatively high absorptivity of dyes, these detection levels can be achieved by using a visible wavelength detector and a column with an internal diameter of 3.2 mm. By employing a column of this diameter as opposed to the standard 4.6 mm, less dilution of the sample occurs and can effectively increase sensitivity by a factor of between two and three. Microbore columns, i.e. columns with diameters of smaller than 2 mm are not favoured for routine analyses because of difficulties in maintaining their efficiency and reproducibility.

There is a trend in the HPLC analysis of dyes to move towards the use of multi-wavelength detectors. The ability to produce simultaneous multi-wavelength plots is a very useful feature of this type of detector because qualitative comparisons between a control and suspect sample are often performed in forensic examinations. An example which illustrates the use of multi-wavelength plots is illustrated in Figure 10.14 where the extracts of dyes originating from 5 mm lengths of a control and suspect fibre have been analysed between 250 and 600 nm at wavelength intervals of 50 nm. The results at 600 nm indicate that the dyes from both fibres are similar. However, a comparison of the plots at the other wavelengths indicates that the suspect fibre dyes extracted do not match those from the control fibre sample and therefore the suspect and control samples do not originate from a common source. Apart from this information, the multi-wavelength plot of the suspect fibre dye extract also show that two other dye components are present. The component eluted at 8.3 min produces a visible λ_{max} of about 450 nm and therefore indicative of an orange-red dye, where as the component eluted at 3.5 min is a yellow dye because it displays a visible λ_{max} at about 350 nm.

Apart from multi-wavelength plots, the numerical multi-wavelength detection techniques of absorbance ratioing and peak purity parameters mentioned previously have also featured strongly in the analysis of dyes. Recently, another numerical method which generates chromaticity coordinates has been introduced and the data generated can enhance sample discrimination. Chromaticity coordinates have been used by spectroscopists for over 50 years as a means of defining mathematically the colour of an object. The x, y and z coordinate values are computed from spectral data and measure the amounts of the three primary additive colours red, blue and green present in a coloured sample as observed by the human eye.

10.3.5 Miscellaneous applications

In forensic laboratories a considerable number of the above methods are in routine use but there are also some applications which are only called upon occasionally. For example, the analyses of sugars, optical brighteners or explosives may be required and methods have been developed for these analytes.

The generally accepted method for separating sugars is to use an amino-bonded silica. However, an alternative method, whereby separations are achieved on a silica column dynamically modified with a long chain amine or a polyethylene amine is preferred since this system is far less susceptible to column 'poisoning'.

Optical brighteners are used to make paper, fibres and other materials whiter and brighter in appearance. These compounds are neutral or ionic and their retention on an HPLC column can be obtained by either reversed phase or ion-pair chromatography. Optical brighteners are naturally highly fluorescent and therefore trace levels can be detected by monitoring the eluate fluorescence. Extreme caution must be taken with these analyses to ensure that any glassware, including new vials produce a blank. This procedure is essential because optical brighteners are used in detergents and if the procedure is omitted, false positive results could be generated.

Finally, HPLC is used for the analyses of explosives and in some forensic laboratories it is being introduced as a routine procedure. Chromatographic separation of explosives can be achieved by reversed-phase chromatography on alkyl-bonded silicas but a major problem has been in finding a technique capable of detecting nanogram quantities of materials. To date, the most sensitive method has been reductive electrochemical detection and provided fairly extensive operational precautions are taken (in particular the exclusion of oxygen), the required detection levels can be achieved.

10.3.6 Conclusions

HPLC is an important analytical technique in a forensic science laboratory since it can be employed for the analysis of an extensive range of analytes in many varied and complex matrices. The potential of this technique is due to its versatility and through the exploitation of separation and detection methods the desired selectivity and/or sensitivity can be achieved only through the development of new techniques or the modification of existing chromatographic techniques. Future applications will still rely on research work performed in forensic laboratories. This is because of the complex and often diverse nature of casework problems which are rarely encountered in other types of analytical laboratories.

Finally, to extend the use of HPLC much further requires detection techniques which can offer still higher sensitivities and improved sample characterisation and discrimination.

Bibliography

Caddy, B. (1991) In *The Analysis of Drugs of Abuse*, ed. T.A. Gough, Wiley, Chichester.
Chakraborty, J., Smith, B., Stoll, M.S. and Perrett, D. (1987) *A Guide to Gas–Liquid and High Performance Liquid Chromatography*, eds. G.S. Challand and M.R. Holland, Association of Clinical Biochemists, London, 1987.
Ho, M.H., ed. (1990) *Analytical Methods in Forensic Chemistry*, Ellis Horwood, Chichester.
Kabra, P.M. and Marton, L.J. (1982) *Liquid Chromatography in Clinical Analysis*, Humana Press, New Jersey.
Kabra, P.M. and Marton, L.J. (1984) *Clinical Liquid Chromatography*, Vols. I and II, CRC Press, Boca Raton, FL.
Lim, C.K. (1986) *HPLC of Small Molecules, A Practical Approach*, IRL Press, Oxford.
Oliver, R.W.A. (1988) *HPLC of Large Molecules, A Practical Approach*, IRL Press, Oxford.
White, P.C. and Tebbett, I.R. (1992) In *The Forensic Examination of Fibres*, ed. J. Robinson, Ellis Horwood, Chichester.

11 Environmental analysis

P.J. RENNIE

11.1 Introduction

High performance liquid chromatography (HPLC) has been widely used for many years in industrial laboratories but its use in environmental laboratories has usually been restricted to analyses such as the determination of polyaromatic hydrocarbons and linear alkylbenzene sulphonates. Traditionally gas chromatography (GC) has been the first choice technique and HPLC only used when GC has proved unsuitable, due to thermal lability or other reasons. This reliance on GC is despite the fact it has been reported that 80–90% of the total organic carbon content in waters is non-volatile and not amenable to GC. Probably the reason for the lack of use of HPLC lies in the poor sensitivity of its most common detector (UV spectrophotometric) compared with GC detectors and the often demanding limits of detection required for environmental analysis, where sub-μg l^{-1} limits of detection are the norm.

Improved UV spectrophotometric detectors, together with automated (or on-line) sample enrichment and/or derivatisation and the development of low cost and highly sensitive detectors such as fluorescence and electrochemical (amperometric and coulometric) systems has meant that HPLC is becoming an increasingly attractive technique, particularly as the development of HPLC has been parallelled by an increasing use of chemicals such as pesticides that are designed to be short-lived in the environment; this readiness to degrade often precludes the use of GC techniques as many of the pesticides nowadays are thermally liable.

The majority of samples dealt with by a typical environmental laboratory are likely to be waters, whether this be groundwaters, rivers, marine or effluents and accordingly the use of reversed phase HPLC has obvious advantages. A look at current literature will show an almost exclusive use of reversed-phase HPLC techniques used for environmental samples.

One of the great advantages of HPLC over other techniques, such as gas chromatography, is its ease of automation and this together with the almost universal scope of UV detection has led to its use in monitoring 'at risk' river waters for organic pollutants in unmanned monitoring stations on two rivers in the UK, i.e. the river Dee in the Grampian region and its namesake in North Wales.

11.2 HPLC mode

Up until the mid-70s HPLC was more or less restricted to normal phase using polar column packing materials. Although some of the organochlorine pesticides separate well using normal phase, the use of normal phase is rarely encountered in the environmental laboratory today, which, as already indicated, is dominated by reversed-phase liquid chromatography. Reversed-phase liquid chromatography has had the most significant impact on the environment laboratory. Highly polar compounds which were difficult to extract from the sample matrices can now often be injected directly or trace enrichment used. Another technique, one that is increasingly important is ion chromatography, where ions in solution are separated using ion-exchange resins. Since the 1970s the use of ion chromatography has increased and is particularly important when low level analysis is required on a multiplicity of ions. Normally dual column systems employing 'suppression' are used. The technique is not suitable for large numbers of samples as analysis time is considerably longer than that when using auto-analyser techniques.

11.3 Sample preparation

In any analysis employing HPLC, sample preparation is as important a stage as the choice of column and mobile phase. Whereas in gas chromatography the choice of extraction solvent in liquid–liquid extraction is seldom limited by chromatographic considerations, in HPLC it is vital that the determinand of interest is compatible both with the mobile phase, the stationary phase and the detector. The sample preparation stage is also important for concentration of the sample and/or derivatising to enhance the detector response. In environmental analysis the sample preparation is generally limited to trace enrichment (including solid phase extraction), liquid–liquid extraction and derivatisation or some combination of these. In some cases, it is even possible to inject the sample directly with no pretreatment. Choice of technique will usually depend on previous work carried out for that particular analyte in that particular matrix. For many solid matrices such as biological samples, soils, etc., supercritical fluid extraction (SFE) is increasingly being used. Off the shelf SFE systems are available which allow for extraction of samples with supercritical fluids, for example supercritical CO_2 resulting in rapid extraction times and avoiding the use of toxic solvents.

11.3.1 Trace enrichment

This is normally carried out off-line using commercially available bonded-phase extraction cartridges such as Bond-Elut, Sep-Pac and Baker-Bond.

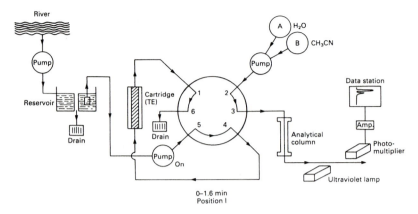

Figure 11.1 Concentration step; river water pumped through trace enrichment column.

The analyte of interest is normally eluted using a small volume of polar solvent such as methanol. This eluant is then injected via a sample loop into the HPLC column. Many packings are now available with more being developed all the time. The technique can be extended to cartridges being stacked and sequential elution of analytes with a multiplicity of solvents. High purity glass cartridges with stainless steel frits are now available for ultra-trace work. Samples can be pulled through under vacuum or forced through under pressure via a pump or syringe. This latter technique could be useful for sampling in the field, the samples would be presented to the laboratory as cartridges which would then be eluted in the laboratory. Concentration and extraction techniques can also be carried out on-line. One way is to automate the bonded phase extraction system using a commercially available system. Eluants are automatically presented to the HPLC without operator intervention. This technique is useful in analysing large numbers of samples. If a single source is to be analysed (say an effluent or river), this can be done more simply using a system shown in Figure 11.1.

Such systems have been used on rivers in the UK to give a 24-h a day monitoring of phenol. In these systems the sample is pumped periodically (e.g. sample pumped for 1 min every 20 min), the sample passes through the trace enrichment column (normally C18) during this time and the organics are concentrated. When the sample pump stops, the Rheodyne valve is rotated (Figure 11.2) whereupon the mobile phase backflushes the concentrated organics onto the analytical column and then on to the detector. The system will detect most organic compounds that can be trace enriched on C18 and can achieve detection limits in the low micrograms per litre.

Analytical columns can also be used for trace enrichment by the con-

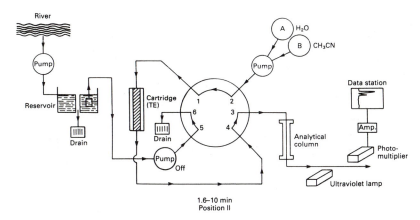

Figure 11.2 Back-flushing step; organic compounds concentrated on trace enrichment column washed on to an analytical column.

centration of non-polar compounds at their head. Very large injection volumes, of 100 ml and more, can be used for preconcentrating. Such a system has been designed for the trace analysis of organic pollutants in water. The unit uses both pre-column trace enrichment and evaporation of the water. UV detection is used to monitor polynuclear aromatic hydrocarbons (PAHs) and 1,1,1-trichloro-2,2-bis-p-chlorophenol DDT and its degradation products.

11.3.2 Direct injection of aqueous samples

Despite the potential for direct aqueous injection of water samples into reverse phase systems, there are very few cases where this is possible due to the low detection levels normally required for environmental analysis. Using direct aqueous injection and coulometric electrochemical detection, the analysis of phenol and chlorophenols and 2-mercaptobenzothiazole have been achieved at trace levels (methods with limits of detection for phenol $0.034\,\mu g\,l^{-1}$ and $0.8\,\mu g\,l^{-1}$ for mercaptobenzothiazole have been achieved). There is a potential for the use of direct aqueous injection for the analysis of phenol in effluents using fluorescence detection which would be expected to detect down to low mg l^{-1}. Direct aqueous injection has been used in an automated system similar to that shown in Figure 11.1. The trace enrichment cartridge was replaced by a large sample loop (50 µl) and a coulometric electrochemical detector used instead of the UV detector.

11.3.3 Liquid–liquid extraction

Liquid–liquid extraction has for many years been the main extraction and concentration technique in the analysis of organics in water. The disadvantage of using this technique for HPLC sample preparation is that non-polar or semi-polar solvents have to be used to avoid miscibility. If reversed-phase HPLC is used any extract will either require re-extraction into a solvent that is compatible with the mobile phase or the extract will require careful evaporation to dryness before redissolving the analytes in mobile phase. This latter technique requires the analytes of interest to have high thermal stability and low volatility. It has however, been successfully used in the analysis of PAHs and is a fairly common approach to PAH analysis in the water industry generally.

One benefit that liquid–liquid extraction has over other techniques is that extraction can often take place in the bottle which was used for sampling, avoiding contamination by transferring the sample and often increasing the extraction recoveries; many pesticides especially the 'drins' (Dieldrin, Endrin and Aldrin) adhere to active sites on the glass walls of the sample bottles. Liquid–liquid extraction of the sample in the sample bottle will increase the recovery for these pesticides over that using other techniques. The problem is often exacerbated when, as is often the case, the sample bottle is cleaned with chromic acid, thereby exposing active sites on the glass walls.

11.3.4 Derivatisation

Due to the selective nature of the most sensitive detectors, electrochemical and fluorescence, derivatisation is often required to take advantage of these. Derivatisation has also been used to improve sensitivity of methods using UV detection by adding a more strongly absorbing chromophore. Derivatisation can be combined with trace enrichment; a detection level of $70\,\mathrm{ng\,l^{-1}}$ for aldicarb (2-methyl-2-(methylthio) propanol O-[(methyl-amino)carbonyl] oxime), aldicarb sulphoxide and aldicarb sulphone has been quoted for 10-ml sample volumes using this technique.

11.3.5 Supercritical fluid extraction

Supercritical fluid extraction (SFE) is becoming increasingly used in the environmental laboratory and is replacing Söxhlet extraction for many determinations. One of the main advantages of using this technique is the speed of extraction. Mass transfer limitations determine the rate at which an extraction can occur. Supercritical fluids have high diffusivities and lower viscosities and thus have much better mass transfer characteristics. It has been reported that quantitative SFEs are generally completed in 10–60 min whereas liquid solvent extraction times can range from several hours to days. The most commonly used supercritical fluid used is supercritical CO_2. Other advantages of using SFE over Söxhlet type extraction are low cost and nontoxicity of the usual supercritical fluids.

11.4 Choice of column

As mentioned previously the most common form of HPLC in the environmental laboratory is that of reversed-phase chromatography. Thus, the most popular column is (as in most other laboratories) C18. Being the most available of HPLC columns, they are inexpensive and available in a wide range of particle size, carbon loading and degree of capping. This could be expected to comprise 95% of LC columns in environmental use. A 15 cm length 6 mm diameter (¼ inch) is probably the optimum (if longer column length is required for separation, then the rest of the chromatography requires investigation). Packings would normally be 5–10 μm with varying capping of unreacted silanol groups depending on the separation. Such a column would be suitable for non-ionic solutes with a molecular weight less than 2000. In certain circumstances such as using a buffered mobile phase (in conjunction with electrochemical detection, say) the pH of the mobile phase may exceed 7. In such cases it is imperative to either purchase a C18 column with pH tolerance or use a divinylbenzene/polystyrene copolymer column.

11.5 Pumping systems

Suitable pumping systems for environmental applications are dual piston reciprocating pumps with at least binary gradient elution with low pressure mixing. This again reflects the trend in other forms of HPLC analysis. With electrochemical detectors it is essential that the pump dampening is adequate to avoid problems and it is often prudent to insert an additional Bourdon tube type pulse dampener comprising a metal expansion coil when using such detection techniques. Multi-solvent pumping systems have one important advantage in the environmental laboratory in that one of the solvents can be used to flush pump and columns out. The reason this is important is that in this type of laboratory the number of LC systems will be small and each system may be required to run varied analyses, say a buffered mobile phase one day for use with electrochemical detection and a normal methanol/water mix the following day for some other analysis. The ability to flush out pumping systems will obviate the need to dissemble and clean up manually.

11.6 Detectors

Whilst there are a large number of types of detector available, the most commonly used in the environmental laboratory will be UV (including diode array), fluorescence and electrochemical (amperometric and coulometric). LC-MS which will hopefully be able to overcome its inter-facing problems has not yet come of age in the environmental laboratory and is available in very few labs.

11.6.1 Fluorescence

With sensitivity and a high degree of selectivity (to reduce interferences) being of importance in the selection of a detector, the first choice must be fluorescence. Unfortunately the selectivity of this detector characterised by exceptionally quiet baseline is its weakness as well as its strength; only 1 in 40 000 compounds fluoresces to any significant effect. Many fluoro-genic reagents are available and can be used in the pre- and post-derivati-sation of pollutants. One area is in the analysis of pollutants resulting from spills of dairy products into river water used for abstraction and future potable use. Phenylacetaldehyde is suspected of being formed from the chlorination of phenylalanine. Phenylacetaldehyde is commonly known as hyacinthin and is very pungent which could result in an off flavour in the water. Analysis by traditional means for trace organic pollutants (i.e. liquid–liquid extraction followed by GC-FID or GC-MS) will not pick up the amino acids; however, a method has been adapted from the clinical analysis of amino acids. Pre-column derivatisation of the sample with o-phthalaldehyde and HPLC with fluorescence detection will allow analysis down to low microgram levels of the amino acid in river water.

The analysis of PAHs by fluorescence detection HPLC is often the water analysts first introduction to HPLC. The analysis of the WHO six PAHs (section 11.8.1.1) namely: fluoranthene, benzo[b]fluoranthene, benzo[k]fluoranthene, benzo[a]pyrene, benzo[ghi]perylene and indeno[1,2,3-cd]pyrene was carried out using fixed excitation (E_x) and emission (E_m) wavelengths. With the advent of relatively cheap variable wavelength programmable fluorescence detectors, the detectors can be optimised for each separate PAH with a resultant lowering of detection limit. Ultratrace determination of PAHs down to 180 fg of benzo[a]pyrene was reported as early as 1983.

Fluorescence could well be used in the on-line system described earlier for the analysis of phenol. Although phenol only weakly fluoresces, one would expect a 10-fold increase in sensitivity over UV detection.

11.6.2 UV detectors (including diode array)

11.6.2.1 Fixed and variable wavelength UV detectors. This is probably the HPLC detector found in most labs. Configurations vary from fixed wavelength (commonly 254 nm) to multi-wavelength and stop run scanning. Its disadvantage as far as the environmental chemist is concerned is the lack of sensitivity. Some form of trace enrichment or derivatisation (as discussed earlier) is almost inevitable. Its big advantage is ease of use, simplicity and, although not a universal response detector, it can be used to estimate unknowns.

11.6.2.2 UV/vis diode array detector. This detector has come into increasing use in the environmental laboratory during the last 5 years. The sensitivity of most modern forms of this detector are comparable with that achieved with a single wavelength system. Identification of an eluting compound is far from definitive unless the compound has an unusual spectra. Its real strength lies in its ability to identify the purity of a peak and so avoiding false quantitation of co-eluting peaks. This ability to detect co-eluting peaks through ratioing of peak absorbances at different wavelengths and observing the tops of the replotted values (a perfect flat plateau indicates high purity) is only shared by MS detection systems. In many diode array systems the wavelength range covered is from 190 to 600 nm and more with a wavelength resolution of 1–2 nm. The light source is normally a deuterium lamp with some systems employing tungsten lamps to avoid the loss of output above 600 nm.

11.6.3 Electrochemical detectors (amperometric and coulometric)

The electrochemical detector has great potential owing to its high sensitivity and is increasingly finding a role in the environmental laboratory. Its selectivity is somewhat less than that of fluorescence detectors. Its principal disadvantage as far as amperometric systems go is the need for frequent recalibration of the detector as the electrodes are gradually 'poisoned'. Coulometric detectors often have an extremely vast surface area to the electrodes which means that the response remains constant until the detector becomes heavily polluted. This obviates the need for constant recalibration. Two other advantages of the coulometric detector are: (i) the response to a particular compound can be predicted using Faraday's equation providing one knows the molecular weight of the compound and the number of electrons transferred during the oxidation (or reduction) (response is inversely proportional to molecular weight and directly proportional to the number of electrons transferred); (ii) two electrodes can be connected in series; the first electrode could be used to reduce an analyte, whilst the second electrode could be used for oxidation. This would restrict the response to those compounds that formed a redox couple (not a very common phenomenon) and could be used to positively identify the determinand; and (iii) the mobile phase can be recirculated as all of the electrochemically active species (at the potentials used) will have been oxidised. If used in a continuous monitoring made in an unmanned river protection station savings from recycling solvent could well run into many thousands of pounds per year, per instrument.

11.7 Sample introduction

With the exception of on-line trace enrichment and/or derivatisation systems, the form of injection is almost exclusively by use of a sample loop with rotary valve (Rheodyne or similar). Normally the size of sample loop is used to decide sample size rather than injection volume. As in other types of laboratory, septum type injection systems in environmental laboratories have long been superseded.

11.8 Analytes

The range of analytes now determined by HPLC in environmental samples is increasing all the time; some of the established techniques used are as follows.

11.8.1 Organic analytes

11.8.1.1 Polyaromatic hydrocarbons (PAHs). PAHs occur in both natural and waste waters. They can be formed as a result of natural decomposition or combustion of organic materials such as wood. Several of the PAHs are known to be carcinogenic and the World Health Organization (WHO) has set a $200 \, ng \, l^{-1}$ maximum acceptable level for the total amount of six indicator PAHs in potable water. Fortunately for the analyst, most PAHs fluoresce strongly and separate well with reversed-phase HPLC. Thus the determination of PAHs is often the first introduction the environmental analyst has with HPLC. Analysis is normally liquid–liquid extraction (dichloromethane is commonly used) followed by careful evaporation of the extract and dissolution in a solvent compatible with the mobile phase (commonly methanol and acetonitrile). Chromatography is carried out using isocratic or gradient elution on a C18 column followed by fluorescence detection. The detector is either of fixed wavelengths or more often wavelength programmable in order to optimise excitation and emission detection wavelengths for each of the determinands. Increasingly the liquid–liquid extraction is being replaced by solid phase extraction with elution using a polar solvent. Detection limits are in the low $ng \, l^{-1}$ region (Figure 11.3).

Although reversed-phase is the norm for PAH analysis, normal-phase LC has been used for the separation of PAHs using phthalimidopropyl-trichlorosilane treated microparticulate silica.

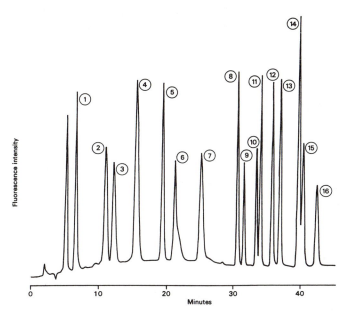

Figure 11.3 Chromatogram of 16 PAH standards using gradient elution (50% aqueous acet-onitrile to 100% acetonitrile). 1, Naphthalene; 2, Acenaphthene; 3, Fluorene; 4, Phenan-threne; 5, Anthracene; 6, Fluoranthene; 7, Pyrene; 8, Benz[*a*]anthracene; 9, Chrysene; 10, Benzo[ ]pyrene; 11, Benzo[*b*]fluoranthene; 12, Benzo[*k*]fluoranthene; 13, Benzo[ ]pyrene; 14, Dibenz[*a,h*]anthracene; 15, Benzo[*ghi*]perylene; 16, Indeno(123cd)pyrene; 17, Perylene.

11.8.1.2 Phenols. Phenols are important environmentally as they are toxic to many aquatic organisms and if the water is a drinking water resource, such as a reservoir, lake or river used for abstraction, even trace (sub-μg l^{-1}) phenols could give rise to taste problems, upon chlorination (and the subsequent formation of chlorophenols), during treatment. The determination of monohydric phenol, being highly water soluble, is tradi-tionally carried out by derivatisation in the aqueous phase followed by extraction of the derivative and analysis by gas chromatography (often electron capturing derivatives would be chosen and gas chromatography with electron capture detection used) or if chromogenic reagents have been used by spectrophotometry. Without derivatisation liquid–liquid extraction would result in very low yields; typically less than 10% due to the hydrophilic nature of phenol. With HPLC solid phase extraction (trace enrichment) either on- or off-line or direct aqueous injection can be used thus avoiding losses due to extraction. The on-line trace enrichment system shown in Figures 11.1 and 11.2 achieves a detection limit of $0.9\,\mu g\,l^{-1}$ in a laboratory-based system and $5.4\,\mu g\,l^{-1}$ in a field-based system (unmanned river monitoring station) for an 8-ml sample of river water using UV detection (270 nm). If fluorescence detection is used then an approximate tenfold increase in sensitivity could be expected (phenol fluoresces only weakly). With the use of electrochemical detectors (phenol is electrochemically active) trace enrichment is no longer necessary and direct aqueous injection can give detection limits in the ng l^{-1} range. Using a 100 μl sample loop and a coulometrically efficient electrochemical detector a detection limit of less than 50 ng l^{-1} has been achieved for a direct aqueous injection of sample. The majority of phenols, cresols, xylenols and the respective chloro-compounds so important in environmental analysis are electrochemically active.

11.8.1.3 Pesticides. Pesticides (insecticides, herbicides, etc.), which are such an important part of twentieth century agricultural practice, are amongst the most closely monitored groups of compounds. Organochlorine pesticides (e.g. Lindane, DDT, Dieldrin, etc.) are normally monitored using liquid–liquid extraction followed by gas chromatography with electron capture detection. These organochlorine pesticides (often highly toxic and persistent in the environment) are being replaced by pesticides more specific in their action and degradable such as organo-phosphorus (e.g. glyphosate) and organo-nitrogen (e.g. chloro-s-triazines) type pesticides. Often compounds that degrade readily in the environment are thermally labile or degrade on the column. In such cases, HPLC provides an effective technique. Usually extraction and concentration is by the use of solid phase extraction, separation is carried out using reversed phase and a C18 column. Detection is dependent upon the characteristics of the compound determined and the limit of detection required. Ideally fluorescence would be the first choice of detection and fluorogenic labelling, either pre- or post-column, could be used for those compounds that do not fluoresce.

One problem that exists in the use of HPLC for pesticide analysis is the difficulty of confirmation of identity. When GC is used for pesticide analysis the sample extracts are normally 'screened' using electron capture (in the case of organochlorine pesticides), nitrogen phosphorus detection (for organo nitrogen or phosphorus pesticides) or flame photometric detection (for sulphur or phosphorus containing pesticides). If a pesticide is detected the extract is then subjected to confirmation by gas chromatography mass spectrometry using positive ion electron ionisation or in some cases negative ion chemical ionisation (e.g. for lindane). This confirmation option is not available when using HPLC as often the analytical technique has been chosen because the compounds will not 'run' on GC, and at present LC-MS lacks the sensitivity for such confirmations at trace levels. The use of more selective HPLC detectors such as electrochemical or fluorescence detectors in correct determination of identity could increase.

11.8.1.4 Linear alkylbenzenesulphonates (LAS). Surfactants in the environment have resulted in foaming in a number of rivers. Of the surfactants used today the most commonly occurring group are the anionic detergents typified by the linear alkylbenzenesulphonates (LAS). LAS can be monitored in our rivers by HPLC using fluorescence detection. The LAS can be extracted and concentrated using SPE followed by separation on a C18 column followed by fluorescence detection E_x225 and E_m275.

11.8.1.5 Organic anions. Carboxylic ions such as formate, acetate and propionate ions, etc. which would normally be difficult to analyse using conventional analytical techniques can be analysed successfully using ion chromatographic exclusion techniques. Using conductimetric detection, sub-ppm detection levels can be achieved. A major advantage of this type of analysis is that a complete series of anions can be analysed simultaneously.

11.8.2 Inorganic analytes

11.8.2.1 Inorganic anions. Although principally associated with organic compound analysis, HPLC has been successfully applied to the analysis of inorganic compounds. The most commonly used form of LC for inorganic analysis is the use of ion chromatography (IC) for the analysis of both anions and cations (although it is fairly rare for cation analysis to be carried out using this technique). Normally conductimetric detection is used (e.g. in the analysis of sulphate, nitrate and chloride). However the use of electrochemical detection has allowed the analysis of anions with a pK_a more than 7 such as cyanide. Detection limits using conductimetric detection are low to sub-ppm (mg l^{-1}) levels and with electrochemical detection low ppb (μg l^{-1}) levels. The advantages of this technique over conventional analyses for these compounds is that a number of anions or cations can be determined in a single run, the method is sensitive and very small (low μl) sample volumes can be used. If a concentrator column is used (effectively trace enrichment) then larger volumes of samples can be concentrated down and low detection levels achieved. The disadvantages of ion chromatography for this type of analyte are that it is time-consuming and other techniques such as auto-analyser or auto-titration systems will allow for a greater throughput of samples.

Nitrate levels in environmental samples can be determined using reversed-phase HPLC with a phosphate buffer as the mobile phase and UV detection. A detection limit of $0.7\,\mu$g$\,l^{-1}$ has been reported.

11.8.2.2 Metals. Although metals analysis has traditionally been carried out using atomic absorption spectrophotometry and more recently inductively coupled mass spectrometry, HPLC can be used in certain circumstances by chelation of the appropriate metal. It would be unlikely that HPLC would be a first choice technique unless it helped to determine speciation. Analysis of arsenic with speciation has been carried out by coupling of an HPLC system to graphite furnace atomic absorption.

11.9 Miscellaneous

Whilst the object of this chapter has been to show the extent and type of HPLC technique that is used today in today's environmental laboratories, there are a number of less routine techniques that may or may not have an impact on routine environmental monitoring. One of the most potentially important of these is the use of LC-MS. The problems associated with using LC-MS for trace analysis are twofold: one is the usual LC-MS problem of interfacing; the second is that of sensitivity of detector. The interfacing problem may well continue to have partial (compared with GC-MS interfacing) solutions such as FAB, and thermospray, etc. However, even given the advances arising from electrospray interfaces the answer may well be to move away from LC-MS to supercritical fluids and SFC-MS.

11.10 Suggested protocol

The flow diagram in Figure 10.4 is intended as a guide and is the way the author would normally approach a new HPLC analysis. Reversed-phase chromatography is assumed and this will mean evaporation of solvent and dissolution in mobile phase if using the liquid–liquid extraction path. No mention has been made of direct aqueous injection as the times that this technique can be employed in environmental analysis are few indeed. It can be seen that the author's choice of detector is fluorescence then electrochemical then UV.

11.11 Conclusions

HPLC has yet to be fully exploited in the environmental laboratory, but when it is, it must inevitably take over the mantle of main chromatographic technique from gas chromatography. The versatility of HPLC encompasses not only organic compounds but organometallic compounds and inorganic compounds. Molecular weight is no longer a barrier nor is polarity or whether ionic or non-ionic. Choice of mobile phase can cover the whole gamut of polarity from hexane to water and a multiplicity of stationary phases. The trend towards highly automated round the clock environmental laboratories will be well served by HPLC with its ease of automation.

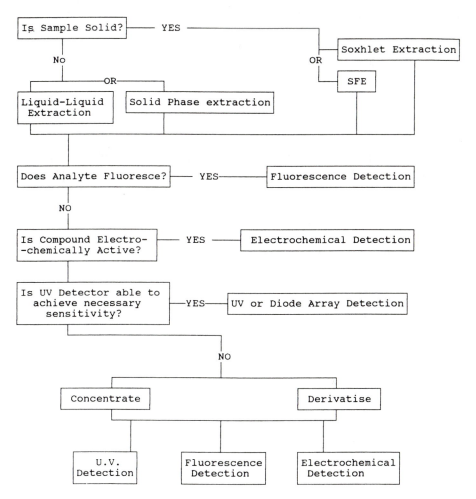

Figure 11.4 How to approach a new HPLC analysis.

Bibliography

Crompton, T.R. (1985) *Determination of Organic Substances in Water*, Volume 1, Wiley, Chichester, 560 pp.

HMSO (1983) High performance liquid chromatography, ion chromatography, thin layer and column chromatography of water samples. *Methods for the Examination of Waters and Associated Materials*, HMSO, London, 99 pp.

Hunt, D.T.E. and Wilson, A.L. (1990) *The Chemical Analysis of Water: General Principles and Techniques*, 2nd edn., Royal Society of Chemistry, Cambridge, 683 pp.

Lindsay, S. (1990) High performance liquid chromatography. *Analytical Chemistry by Open Learning*, Wiley, Chichester, 244 pp.

12 Food, organic and pharmaceutical applications

W.J. LOUGH and I.W. WAINER

12.1 Introduction

This chapter provides a 'catch-all' covering applications for which no or minimal sample preparation is required. More complex applications (forensic, biomedical and environmental) for which sample preparation are often a critical step are dealt with in Chapters 10 and 11.

By dealing with food, organic and pharmaceutical applications simultaneously the comparisons and contrasts between them may be seen. Generally, the types of compounds involved are low molecular mass organic molecules and these need to be analysed in samples of the compound itself (bulk substance) or in a mixture (product or formulation). However, for convenience, there is also a brief mention here of large molecular mass compounds as drugs or as food constituents. Synthetic organic high molecular mass compounds (polymers) are dealt with in Chapter 9.

12.2 Bulk substances

12.2.1 Monitoring reactions

Liquid chromatography (LC) is not only used in the analysis of bulk substances but also in monitoring the reactions used to prepare them. The simplest form of this is when the disappearance of starting material is followed to determine the point at which the reaction is complete. Normally the appearance of product(s) would also be monitored. For laboratory scale synthesis, particularly when preparing new chemical entities, it is thin-layer chromatography (TLC) and not LC that is the most suitable technique. TLC is cheap, simple and perfectly adequate for this non-demanding type of analysis. Importantly, using TLC it is also possible to detect all components of a reaction mixture. This is not the case for LC since some components may have a poor ultraviolet (UV) chromophore (if UV detection is being used) and/or be completely retained on the LC column.

Given this, there are instances when it is necessary to use LC. If the reaction mixture is very complex and it is desired to follow the appearance and disappearance of by-products, the higher efficiency of LC is useful.

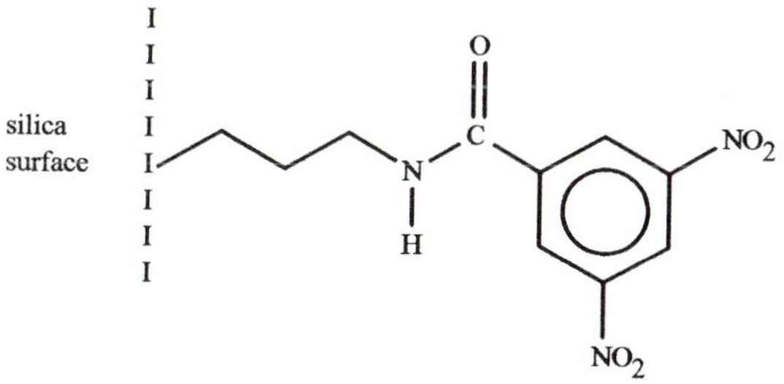

Figure 12.1 Dinitrobenzoylated-aminopropyl-silica. This phase is prepared by a catalysed reaction of 3,5-dinitrobenzoic acid with aminopropyl-silica. Retention takes place through charge transfer interactions when an analyte containing an electron-rich π-electron system interacts with the electron-deficient dinitrobenzoyl- system.

One such example is the $PdCl_2/V_2O_5$ catalysed conversion of nitrobenzene to phenyl isocyanate at high temperatures and under high pressures of carbon monoxide;

$$PhNO_2 + CO \longrightarrow PhNCO + CO_2 \qquad (12.1)$$

The starting material and product may be determined by gas chromatography (GC) but the (many) by-products are involatile. It is therefore best to monitor the reaction by LC. The reaction may be sampled even at high pressure by using a stainless steel capillary. The reactive isocyanate is converted to a urethane when the sample is added to ethanol. This ethanolic solution may then be analysed using straight phase LC on silica. Since the reaction components have a wide range of polarities, two different mobile phases are required to determine all of them. However if an alternative stationary phase is used (Figure 12.1) the complete analysis may be carried out with a single isocratic mobile phase (i.e. composition does not change during the run). The separation on the dinitrobenzoylated-aminopropyl-silica stationary phase of the principal components usually present in the ethanolic solution is shown in Figure 12.2.

When such an LC system as above is available to monitor a reaction it may also be used quantitatively to gain information on the reaction kinetics and possible mechanism. This is very important for industrial processes. Hence LC may be used for this purpose and to optimise reaction conditions even when studying much simpler reactions.

12.2.2 Synthetic intermediates

The determination of the purity of intermediates in a synthetic route to a target compound is more common in pharmaceutical analysis than organic or food analysis. This is because lengthy syntheses are more common in the preparation of drugs than foods or general organic compounds.

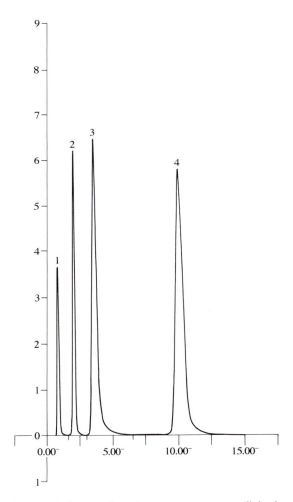

Figure 12.2 LC of carbonylation reaction mixture components on dinitrobenzoylated-amino-propyl-silica (150 × 4.6 mm i.d.); mobile phase *n*-hexane: propan-2-ol (60:40, v/v); flow rate 1.0 ml min⁻¹. Peak 1; phenyl ethyl carbamate. Peak 2; N-phenylcarbamylbenzoxazol-2-one. Peak 3; *sym*-N-diphenylurea. Peak 4; triphenylbiuret. On silica *sym*-diphenylurea is much more strongly retained than the other three components.

Constituents of foods are frequently natural and not synthetic. Synthesis of organic compounds for general use tend not to be as complicated. Drug synthesis is frequently lengthy since only a very specific molecular shape and charge distribution will give the desired pharmacological effect and this is often not available from a simple, easily accessible compound.

The determination of the purity of drug intermediates becomes more important in the later stages of the drug development process. By this time reference standards of the intermediates (samples of known, high purity) are available and are used in main peak relative assays (Chapter 7) to determine the purity of samples of the intermediate. In other words the purity is determined by using peak areas to compare with the purity of the reference standard.

The purity of the intermediate is of interest in that it has a bearing on the subsequent purity of the final target drug and, more importantly, on the overall yield of the synthetic process, i.e. each step will be more successful if its starting material is pure. Thus, important though the intermediate purity is, it is not as important as the purity of the drug since the intermediate is not administered to man. Consequently the methods used are the same as for the drugs themselves but there is greater flexibility in the acceptance criteria attached to the method validation tests. For example, while an RSD of <0.5% might be required for the reproducibility of an LC assay for a drug, 2–3% might be acceptable for a drug intermediate.

12.2.3 Purity

The method used for the LC determination of the purity of a bulk substance will depend on circumstances.

12.2.3.1 Rapid screen. In the early stages of drug discovery thousands of analogous compounds may be prepared in the study of a therapeutical area. These compounds are subject to pharmacological testing and are of further interest only if there is significant pharmacological activity. Purity is not of paramount importance since pharmacological activity which is much greater than existing drugs is usually being sought and if a compound has such activity it will be picked up with even a relatively impure sample. Further, the fortuitous situation may arise that an impurity gives pharmacological activity. Given these circumstances extended method development for any one drug candidate is not warranted. A rapid generic method that will cope with large numbers of samples would be more useful.

The nature of the LC conditions adopted for such a rapid generic LC method would depend on whether the compounds were acidic, neutral or basic and whether they were hydrophilic or hydrophobic. Circumstances however, would favour the use of 'fast LC' i.e. the use of short columns packed with 3 μm particles and operated at high flow rates. Using fast LC, efficiency is satisfactory and compounds having very large capacity factors may be eluted within reasonably short times.

12.2.3.2 Main peak assay. Once a drug candidate progresses into its development phase a main peak relative assay for the purity determination of samples is mandatory. As indicated above, and in Chapter 7 acceptance criteria for method validation tests have to be stringent, especially with respect to specificity.

There are different issues involved in the LC purity determination of compounds used in foodstuffs. For example since the compounds themselves do not bring about powerful pharmacological effects, there is less likelihood of structurally-related impurities being highly toxic. In principle, it may be possible to carry out less stringent testing for method specificity. However the analyst needs to be wary of the possible presence of toxic non-structurally-related contaminants present.

With the major constituents in foods the choice of LC detector is often the most important issue. Compounds such as vitamins, carbohydrates etc. may not have a strong ultraviolet (UV) chromophore. Therefore refractive index (RI) detection and, increasingly, electrochemical detection are often used. As discussed later, the choice of detector is even more important when determining the concentration of components in the foodstuff rather than the bulk constituent.

It is also important in food analysis to consider that some common compounds with high molecular mass may be present in foods, such as polysaccharides and proteins. LC of these compounds requires larger particle pore sizes than usual (> 100 Å). They also need to be analysed frequently using modes of LC other than reversed-phase. For proteins in particular there are a range of modes of LC such as size-exclusion, ion-exchange and hydrophobic interaction (Chapter 3) that allow the protein to be analysed in non-denaturing conditions (i.e. the protein retains its three-dimensional structure). While this is vital for preparative work where retention of the initial biological activity of the protein is required, it is not essential for quantitation. LC of the completely denatured protein may be carried out. The important point is that the denaturing process does not take place during the chromatographic process as this will lead to band-broadening.

12.2.4 Impurities

The procedures for the LC determination of impurities in bulk drugs are outlined in Chapter 7. These determinations become more important as a drug candidate gets closer to production and the market. Fortunately, the analysis becomes easier at this stage since reference standards for many of the impurities will have become available. This allows more accurate determination than using methods which assume an unknown impurity to have the same UV molar absorbance as the drug. The level to which drug impurities need to be determined is currently around 0.1%. However this is lowering as safety issues continue to be of paramount importance, and also varies with the anticipated (or known) toxicity of the impurity.

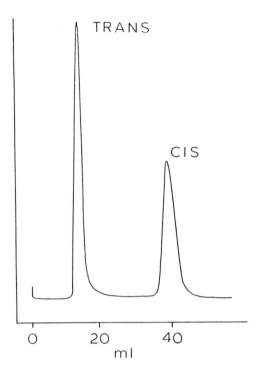

Figure 12.3 Separation of *cis* and *trans* butadienone iron tricarbonyl complexes. Column, 300 × 4 mm i.d. Partisil SI-10 (Whatman Inc.); mobile phase, methylene chloride. (Reproduced from D.G. Gresham *et al.*, 1977.)

As alluded to earlier, the important impurities in food components may well not be structurally related. Accordingly, this presents a difficulty; unless there is prior knowledge of the types of impurity likely to be present and their physicochemical properties, the impurities may go undetected through unsuitable choice of LC conditions and/or detector type.

12.2.5 Isomer separations

Often when a compound exists in different isomeric forms it is used as a mixture of isomers. Alternatively there may be a need to monitor samples of one isomer to ensure its purity. In both cases it is important to be able to separate isomers using LC. This is especially important for drug substances, since isomers will almost certainly have different pharmacological and toxicological properties.

Isomer content is of interest in organic chemistry since a reaction which is part of a synthetic route will have subsequent steps in the synthesis which usually require one isomeric product. Another perspective is that

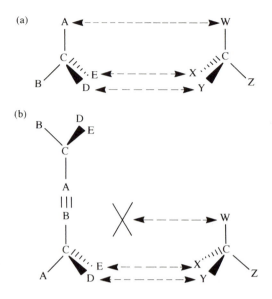

Figure 12.4 Illustration of the 3-point interaction role. All the three interactions which take place in (a) are not possible in situation (b), where the chiral selector approaches the opposite enantiomer.

the isomeric ratio of a reaction product will often give clues to the likely reaction mechanism.

The illustration of an isomeric separation shown in Figure 12.3 is an organometallic example. It is interesting to note that the mode of LC used is straight-phase. This has largely been superceded by reversed-phase LC for many applications, but although it has its disadvantages for polar, basic compounds (as can be seen), it can provide excellent selectivity which allows some very difficult separations to be achieved even with relatively inefficient columns.

Enantiomers are types of isomers that have aroused much interest since the early 1960s when the drastic teratogenic effects of the drug thalidomide (children born to mothers who had been taking thalidomide were born with truncated limbs) were attributed to one of the two enantiomers present in the drug substance. Enantiomers not only have the same molecular formula, they also have the same distribution of bonds in space. They are however non-superimposable mirror images of each other. The possibility of enantiomers existing is a feature of asymmetric molecules and most commonly arises when a molecule possesses an asymmetric carbon. Such a situation is shown in Figure 12.4. Enantiomers have identical physicochemical properties and therefore cannot be separated on

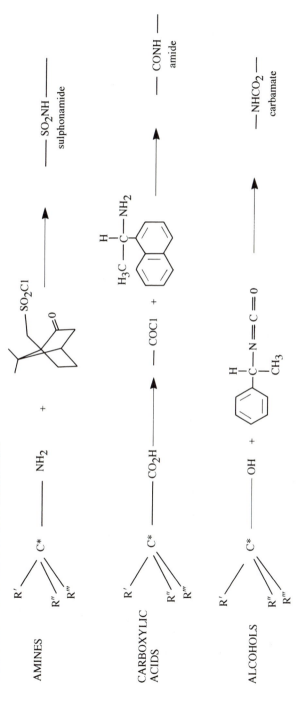

Figure 12.5 Typical common diastereomer formations.

conventional chromatography. It is also shown in Figure 12.4 how enantiomers can only be distinguished when they interact with one of the enantiomers of a molecule which itself is chiral (i.e. able to exist in different enantiomeric forms). Moreover, if the gross oversimplification is made that all interactions are point-to-point interactions, a minimum of three such interactions (attractive or repulsive) are required for the enantiomers to be distinguished. Thus in the LC separation of enantiomers it is necessary to use a 'chiral selector'. This may take the form of a chiral derivatising agent, a chiral mobile phase additive or a chiral stationary phase.

The types of derivatising reagents that may be used are shown in Figure 12.5. The products formed are called diastereomers. These compounds which possess two 'chiral centres' have different physicochemical properties and are therefore separable by LC on conventional 'achiral' LC columns.

For a number of reasons it is usually preferable to use a chiral stationary phase (CSP) rather than a chiral derivatising agent. A CSP will normally be preferred to a chiral mobile phase additive, partly because the use of a chiral selector as a mobile phase additive will lead to much

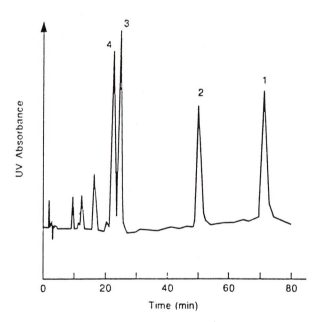

Figure 12.6 Chromatogram of the four stereoisomers of aspartame (APM). Peak 1; L,L-APM. Peak 2; L,D-APM. Peak 3; D,D-APM. Peak 4; D,L-APM. The early eluting peaks are degradation compounds. Using a Crownpak CR(+) column and propan-2-ol aqueous perchloric acid mobile phase, a temperature gradient was required to elute peaks 1 and 2 (Motellier and Wainer, 1990).

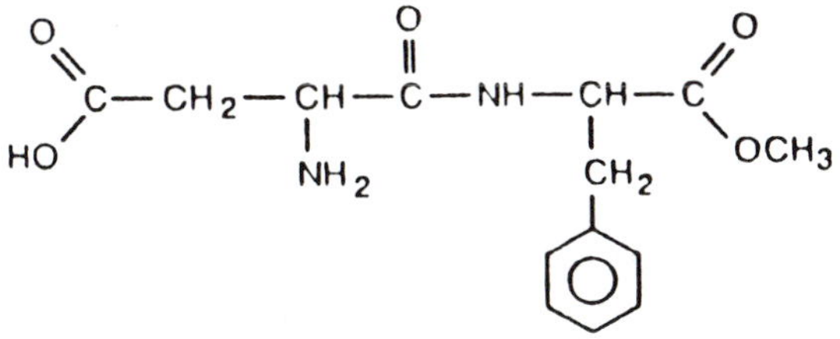

Figure 12.7 Structure of aspartame (APM).

higher consumption of an often expensive compound. Enantiomerically pure chiral compounds occur in nature and are readily available fairly cheaply. These compounds therefore form the basis of most commercially available CSP. The most commonly used are proteins, cyclodextrins (cyclic oligosaccharides), derivatised polysaccharides and derivatised amino acids.

The example of an LC chiral separation shown in Figure 12.6 serves to emphasise (a) that the demand for effective chiral selectors is such that even complex synthetic chiral selectors have been commercialised, and (b) the interest in chirality extends beyond pharmaceutical applications, being widespread and in this instance being found in food analysis. Aspartame (N-DL-α-aspartyl-DL-phenylalanine methyl ester (Figure 12.7)) can exist as four stereoisomers, DD-, LL-, DL- and LD-. On an achiral column DD- and LL- would appear as a single peak which would be separable from another single peak arising from DL- and LD-. A chiral column is needed to separate the enantiomeric pairs (i.e. DD- from LL- and DL- from LD-). The LL-isomer is used as artificial sweetener (under the brand

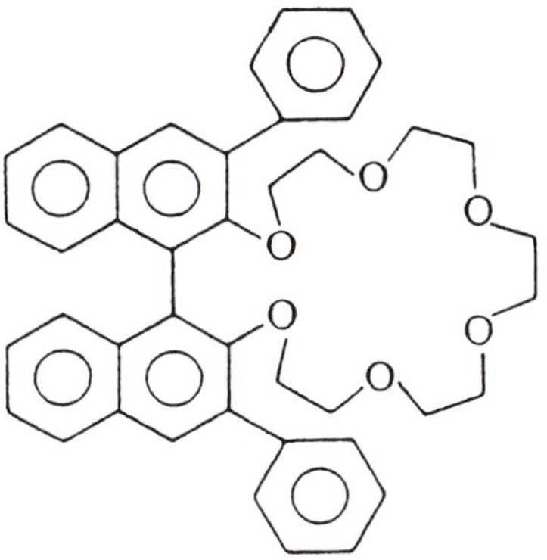

Figure 12.8 Chiral crown ether used in commercial LC chiral stationary phases.

name NutraSweet) and should be free from the other stereoisomers. As illustrated in Figure 12.6 this can be tested by carrying out LC on a Crownpak CSP and possible degradation products may also be detected using the same test conditions. Crownpak is based on the crown ether shown in Figure 12.8. The chirality arises not from an asymmetric carbon but in this case from restricted rotation caused by steric hindrance around the oxygen atoms in the binaphthoxy-moiety.

12.2.6 Preparative isolation

When using LC for preparative isolation rather than for quantitation, one approach is to simply scale up an analytical separation. This is a common approach when the objective is to obtain quantities of pure compounds in the range 10–1000 mg from complex mixtures. The starting point is to increase the load (by increasing mass in the same volume or increasing the volume of the same concentration solution, depending on the circumstances) as much as possible while just retaining baseline resolution of the peaks of interest from the peaks running closest to them. These conditions may then be scaled up using a factor of the ratio of the column volumes. For example, in moving from a 250 mm × 5 mm to a 250 mm × 10 mm column, the ratio is 4 (volume ratio = $(\pi r^2 h)_2/(\pi r^2 h)_1 = (\pi \times 10^2 \times 250)/(\pi \times 5^2 \times 250) = 100/25 = 4$). The concentration of the solution injected should stay the same. However if a 50 µl injection volume were being used then a 200 µl injection volume should be used on the larger column. Similarly a flow rate of 2.0 ml min^{-1} on the analytical column would become 8.0 ml min^{-1} on the larger column. A typical example of a preparative separation developed in this way is shown in Figure 12.9.

When the isolation of larger quantities (i.e. >1 g) of pure material are required and throughput (i.e. mass of pure material isolated per unit time) is a consideration, then other approaches to preparative LC come into their own. Throughput can be improved when using the direct scale-up approach described above by the use of automation. Repeat auto-injectors and intelligent fraction collectors are now commercially available. However if the samples being processed contain only two or three components which are reasonably well resolved a better approach might be to overload a preparative column packed with large particles (20–40 µm). To appreciate why this is an attractive option it is instructive to understand that (i) efficiency drops off with load more sharply for small particle sizes, so that under overload conditions the efficiencies obtained for 5 µm and 20–40 µm particles are fairly similar, and (ii) how few theoretical plates are required in a column if there is good selectivity (Table 12.1).

Clearly there is no advantage to be derived from using small particles if

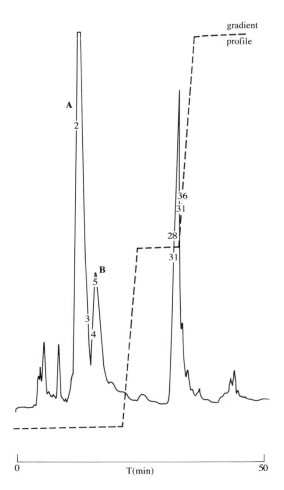

Figure 12.9 Preparative HPLC of β-cell tropin. Column; Lichroprep RP8, 250 × 16 mm. Mobile phase; acetonitrile/water/trifluoracetic acid; 32/68/0.1, followed by gradient step to clean up the column. Detection; UV 274 nm. Flow; 8.0 ml min⁻¹. Injection volume; 2.0 ml. Sample load 35 mg. (a) main peak; (b) major impurity.

the column is being overloaded. From Table 12.1, it can also be seen that the number of theoretical plates available from columns packed with 40 μm particles will be adequate for many separations. If throughput is the highest priority then this overloading approach may be modified by overloading the column well beyond the point where baseline resolution is lost, collecting only parts of the peaks containing a single pure component and saving eluent from parts of the peaks where there is overlap so that these fractions can be subsequently recycled (Figure 12.10). Because of the recycling this may seem to be a time consuming procedure. However, it

Table 12.1   Calculated* N_{min} values for varions α values

α	N_{min}
2	23
1.5	92
1.3	256
1.2	575
1.1	2 300
1.05	9 200
1.03	25 556
1.01	230 000

*When $R_s = 1.0$ and average $k' \approx 5$, using $N_{min} = 23 \, (\alpha - 1)^2$; for preparative work k' values less than 5 are advisable to minimise overload of the stationary phase to increase throughput and to minimise solvent consumption. (Courtesy of Dr. D. Wallworth BASTechnicol Ltd.)

tends to be faster overall because much larger quantities of material are processed in each run.

A further approach to preparative LC is the use of displacement chromatography. With this technique the sample is loaded onto the top of the column and then sample components are sequentially displaced on pumping through a solution containing a displacer compound which has a higher affinity for the stationary phase than the sample components. The displacer itself is then removed from the column by pumping through a strong solvent. The column is then reequilibrated with the initial solvent and the whole process begins again. Using displacement chromatography, very high loads of sample components can be resolved in one 'run'. The main disadvantage is that method development for finding the most suitable displacer can be time-consuming.

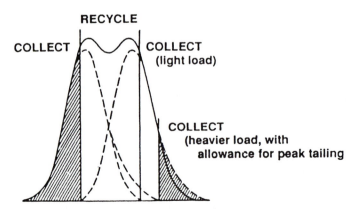

Figure 12.10 Use of recycling in preparative LC with partial resolution.

12.2.7 Determination of physico-chemical parameters

12.2.7.1 Partition coefficients. Knowledge of partition coefficients, usually for the distribution of a compound between octanol and physiological buffer, is of value in pharmaceutical science because it provides information about the lipophilicity of a drug and from this, some idea of how it might act in the body. When partition coefficients are measured directly, (a) it is necessary to use milligram quantities of very pure compound; (b) the experiments are very time-consuming; (c) the experiments must be conducted with great care, and (d) the experimental procedure must be modified to accommodate compounds with very high or very low partition coefficients. It is much more convenient to estimate the partition coefficient (or $\log P$ value) of a compound by using LC. In the past some researchers used liquid–liquid partition LC with octanol coated on a solid support. Such lengths are not necessary as a perfectly adequate correlation exists between $\log P$ and $\log k'$ values for reversed phase LC on octadecylsilyl-silicas that are well end-capped. Within a set of analogues the correlation may be as good as $r = 0.99$. A set of compounds with a known $\log P$ are chromatographed to create a calibration plot which is then used to estimate $\log P$ values from $\log k'$ values of sample compounds. Using such methodology, $\log P$ values can be determined very quickly with sub-milligram quantities of impure samples irrespective of the studied range of $\log P$. Using fast LC (as mentioned earlier, high flow rates with short columns packed with 3 µm particles) the time taken to determine $\log P$ values of a large set of compounds can be reduced even further, as shown in Table 12.2. By carrying out reversed-phase LC over a wide range of pH values it is also possible to use reversed-phase LC to estimate pK_a values. However pK_a values may be determined satisfactorily by other means.

12.2.7.2 Intermolecular interactions. LC can be used in two ways to determine intermolecular interactions; (a) by using one compound in the

Table 12.2 Example sample set of 30 compounds with k' values spread through the range 13–40.

	Total chromatographic run time (h)	
	manual injection	autosampler injection[*]
conventional LC (250 × 5 mm, 2.0 ml min⁻¹)	2.31	25.6
fast LC (50 × 5 mm, 4.0 ml min⁻¹)	0.45	4.9

[*]The next sample must not be injected until k' = 40.

Table 12.3   Chiral resolution of 2-aryl propionic acid non-steroidal anti-inflammatory drugs on the HSA-CSP

Solute	Without octanoic acid		With octanoic acid		
	$^a k_2'$	α	[mM Acid]	$^a k_2'$	α
Naproxen	63.50	1.32	2	15.06	1.15
Flurbiprofen	*	*	4	21.50	1.27
Ibuprofen	61.3	3.51	4	10.12	1.33
Benoxaprofen	59.18	1.16	4	19.90	1.12
Pirprofen	36.32	1.47	0.5	15.41	1.17
Suprofen	34.62	2.51	4	4.90	1.19
Ketoprofen	10.10	1.24	0	–	–
Indoprofen	11.4	1.27	0	–	–
Fenoprofen	26.23	1.60	2	10.76	1.31

$^a k_2'$ = the capacity factor of the more retained enantiomer
*In the absence of octanoic acid, the peaks for flurbiprofen were extremely ill-defined, eluting with capacity factors greater then 80. Reproduced from Nocter *et al*, (1991)

mobile phase at a range of concentrations and carrying out LC of the other compound, or (b) immobilising one molecule and carrying out LC of the other. The latter method is more difficult to set up and is more prone to giving misleading results because of the linking groups used in the immobilisation interfering with the interactions. However, for instance, the phase shown earlier (Figure 12.1) retains compounds through π–π interactions and may be used to estimate the charge transfer properties of analytes by LC. Such information is occasionally used along with $\log P$ and pK_a information in quantitative structure activity relationships to rationalise the pharmacological activity of drugs.

12.2.7.3 In vivo modelling.   While the techniques described in section 12.2.7.2 are not frequently used for small molecules, they are increasingly being used, along with some other more sophisticated LC techniques, to study interactions between small molecules and large molecules, in particular to study drug–protein binding. Wainer and Nocter have been prominent in this field and have used an immobilised human serum albumin (HSA) phase to study the interaction of drugs such as benzodiazepines, warfarin and ibuprofen with HSA. Some of these workers' data is shown in Table 12.3. The k' values of the non-steroidal anti-inflammatory drugs, when the mobile phase modifier is not used, are good indicators of the strength of their binding to HSA.

The use of this phase as a model of interactions that take place *in vivo* may be taken a stage further. By adding a drug to the mobile phase it is possible to study how the interaction of one drug with HSA is affected by the presence of another.

12.3 Products or 'formulations'

12.3.1 Content determination

The determination of its content is the most common analysis carried out on a product or formulation. There are some obvious differences between such determinations and the determination of purity of a bulk substance.

The first difference is that in the product or formulation other compounds are present. If these compounds interfere unduly with the LC method or are present in much larger quantities than the compound of interest then some sample pre-treatment will be necessary. This has been dealt with in Chapter 8. However for a product or formulation this may be straightforward, frequently consisting only of selective dissolution followed by injection of supernatant or a filtration step. The choice of solvent for selective dissolution may be simple but for compounds in a complex formulation (such as a drug in animal diet) it might take some time to find suitable conditions to recover all the compound from the solid residue.

The other difference between bulk substance purity determinations and content determinations is that in the content determination a lower concentration of the compound of interest is being determined. This and the

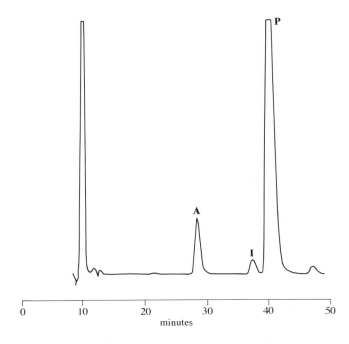

Figure 12.11 Sample chromatogram for separation of pilocarpine (P), isopilocarpine (I) and pilocarpic acid (A). A phenyl-bonded silica was used with mobile phase of acetonitrile-aqueous buffer (3:90 v/v), (pH 2.5).

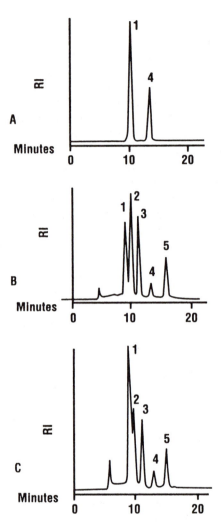

Figure 12.12 Carbohydrate profiles of plain and flavoured yoghurts on the Aminex HPX-87P column, 300 × 7.8 mm. Eluant; H_2O. Flow rate; 0.6 ml min^{-1}. Temperature; 85°C. Detection; RI at 32×. (a) Plain yoghurt, diluted 1:1; 20 μl. (b) Strawberry yoghurt, diluted 1:2; 20 μl. (c) Blueberry yoghurt, diluted 1:1; 20 μl. Peaks: 1, Sucrose (and maltose in (c)); 2, Lactose; 3, Glucose; 4, Galactose; 5, Fructose.

presence of other compounds has an effect on LC method validation. In specificity testing the separation of the compound of interest from other compounds present in the sample (separation of 'active' from 'excipients') must be demonstrated. Because of the lower concentration and the nature of the information sought by a content determination, a higher relative standard deviation on the precision test might be acceptable. It also becomes more relevant to carry out a test for limit of quantitation and limit of detection.

The determination of the content of a drug formulation will often be simpler than the content of a foodstuff. Apart from the usual exception of natural folk medicines used in developing countries (these may rely on the synergistic effect of a mixture of actives) most pharmaceutical formulations contain two actives at most. In fact, the formulation may often be so simple that the determination can be carried out by UV spectrophotometry.

A sample chromatogram for the LC conditions used for the determination of pilocarpine in pilocarpine ophthalmic solutions is shown in Figure 12.11. This is atypical since there is a very long analysis time, due to setting up the method to determine degradation products. However it is typical in that, for commercial samples containing 2%, 1% and 0.5% pilocarpine, RSDs in the range 3.2–4.1% were considered to be excellent.

With respect to the determination of the content of a foodstuff there will usually be a much larger number of constituents. Moreover some of these constituents may be present in quite low concentrations. This exacerbates the detection problems referred to earlier (section 12.2.3.2) with the result that sensitive detectors such as electrochemical detectors, when appropriate, come into their own. As well as the vitamins and carbohydrates mentioned earlier, electrochemical detectors may be used for additives such as antioxidants and antimicrobials. However at the present time, RI detection and UV detection at low wavelengths are the most common means of detection for food components with poor UV chromophores. UV detection at 210 nm is often the preferred option for the detection of organic acids (which may be used to modify flavour or may be analysed to give an indication of product stability), fats and oils. An example of RI detection is shown in Figure 12.12. The analysis of carbohydrate content of dairy products is required for flavour studies and nutritional assessments. In the example shown it can be seen that plain yoghurt contains natural lactose and galactose while strawberry yoghurt to which corn sweetener has been added for flavour enhancement, and blueberry yoghurt which contains added corn syrup, contain additional carbohydrates. The type of column used in this instance is one of a family of columns containing polymeric resins which separate carbohydrates by a combination of size-exclusion and ligand-exchange mechanisms. There is also a need for carbohydrate analysis in ice cream, fruit juice, wines and even cookies and baked potatoes.

The content of contaminants in foods is an altogether more demanding problem since contaminants are often present in trace quantities. Accordingly, it is necessary to resort to the sample preparation and trace enrichment techniques commonly used in environmental analysis. The types of problems that might be encountered include drug residues in meat products, furosine (known for its deteriorative and browning reaction) in dairy products, aflatoxins (carcinogens) in a variety of foods including nuts and dried figs, and hesperidine (from green skin on oranges) in orange juice.

12.3.2 Dissolution studies

LC is used extensively in dissolution studies carried out on pharmaceutical formulations to assess the likely availability of the drug substance from the formulation when it enters the stomach. The formulation is stirred in a dissolution apparatus consisting of a standard stirring device and dissolution bath usually containing an aqueous buffer designed to mimic conditions in the stomach. The aqueous buffer is then sampled over a set time period and analysed for drug concentration. This type of analysis differs from content analysis of formulations in that the initial sample solutions contain a lower concentration of the drug and there are more samples to analyse. Despite the lower concentrations, determination by UV spectrophotometry will often suffice and is used when possible in the interests of speed and simplicity. Given the high sample numbers and the often undemanding separation required, this is an LC application area where the benefits of fast LC may be exploited.

It is also useful to follow the dissolution of excipients since this may control the release of the drug substance. LC may be used for this purpose but such analysis is more difficult since the excipients may have a weak chromophore and/or be polymeric. It goes without saying that simultaneous determination of drug and excipient in dissolution samples is more difficult still.

12.3.3 Stability studies

While stability studies are carried out on foods, the study of stability in pharmaceuticals is even more vital because of the need to avoid potentially toxic degradation products. Accordingly stability studies in the development of formulations are very thorough. Several samples must be studied at each of a wide range of temperature, humidity and light conditions and samples are taken regularly at intervals over total times of up to two years.

In such studies it is necessary to demonstrate that the drug content of the formulation has not changed with time. Also if degradation does take place it will be necessary to identify and quantify degradation products. (It might first be necessary to use TLC to locate any new degradant.) Fortunately little new method development is likely to be needed for this task since in the validation of the LC assay for drug purity or drug content in the formulation it is likely that checks will have been made for resolution from likely degradation products. A good illustration of this is the LC conditions developed for the determination of pilocarpine in ophthalmic

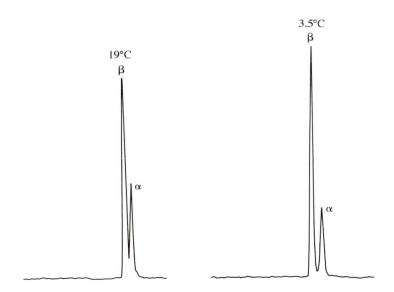

Figure 12.13 Separation of lactose anomers on ODS Hypersil [2 × (250 × 4.6 mm)] i.d. Mobile phase; water. Flow rate; 1.0 ml min⁻¹. Detection; RI.

solutions (Figure 12.11). Identical conditions are used for the determination of pilocarpine degradation products i.e. isopilocarpine and pilocarpic acid.

The study of stability is not restricted to looking at degradation of the 'active'. It is prudent to bear in mind that any degradation of a formulation excipient may lead to a change in the drug release characteristics of the formulation. For example lactose, a frequently used excipient, can undergo anomerisation in solution between its α- and β- forms. It is conceivable that this could take place in the solid state in a formulation stored under extreme conditions. This could be studied by LC (Figure 12.13).

Finally, as a reminder that stability studies are not restricted to pharmaceutical products, a chromatogram taken from the study of a stability study of LL-aspartame (Nutrasweet[T]) in diet cola is shown in Figure 12.14. The same conditions were used for the resolution of the stereoisomers of aspartame (Figure 12.6), and the study of the stability of LL-aspartame in distilled water and coffee sweetened with Nutrasweet[T].

12.4 Conclusion

LC is by far the most frequently used separative technique in food, organic and pharmaceutical analysis. There are some opportunities for

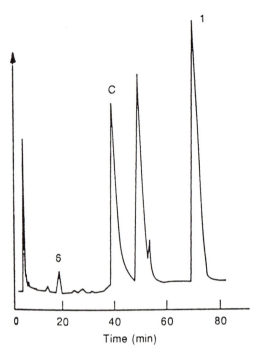

Figure 12.14 Determination of L,L-aspartame and its decomposition products in a diet cola. 1; L,L-aspartame, 6; a diketopiperazine degradation product and C; caffeine. Conditions are as for Figure 12.6. (Motellier and Wainer, 1990.)

GC in the analysis of foods and organics but this is less the case for pharmaceuticals, which by virtue of the nature of their intended use, tend to be polar, ionisable compounds and therefore involatile. TLC is used instead of LC for applications such as; (a) where a very simple cheap method is required, (b) when it is necessary to visualise all components of the sample, and (c) when dealing with high sample numbers (e.g. stability studies). The latter case is an example of when modern TLC instruments that allow quantitation are worth considering. Under other circumstances, the fact that advanced TLC detracts from the simplicity and low cost of the technique results in its perception as a technique complementary to LC rather than a competitor. With respect to pharmaceuticals it is also being said that capillary electrophoresis (CE) is complementary to LC. In fact increasing use of CE is the only thing in the foreseeable future likely to have any impact upon the continuing dominance of LC amongst separative techniques used in pharmaceutical analysis.

Bibliography

Gresham, D.G. *et al.* (1977) *J. Organometallic Chem.* **142** 123–131.
Lough, W.J. (Ed.) (1989) *Chiral Liquid Chromatography* Blackie A&P, Glasgow.
Motellier, S. and Wainer, I.W. (1990) *J. Chromatogr.* **516** 365–373.
Townshend, A. (Ed.) (1995) *Encyclopaedia of Analytical Science*, Harcourt Brace, London.
Wainer, I.W. (1987) Proposals for classification of HPLC chiral stationary phases. *Trends in Anal. Chem.* **6** 125–134.

Index